Navigating the Cyber Maze

In an era where cyber threats loom larger than ever, *Navigating the Cyber Maze: Insights and Humor on the Digital Frontier* offers a refreshing blend of deep insights and engaging humor to demystify the complex world of cybersecurity. Authored by Matthias Muhlert, a seasoned cybersecurity expert with over 20 years of experience, this book aims to provide readers with a comprehensive understanding of cybersecurity, extending far beyond traditional IT concerns. From safeguarding smart homes to securing agricultural supply chains, Muhlert's expertise shines through in this essential guide.

What sets this book apart is its unique approach to making cybersecurity accessible and enjoyable. Matthias not only breaks down intricate concepts with clarity but also infuses humor throughout, making the learning experience both informative and entertaining. Whether you are a seasoned professional or new to the field, this book ensures you will gain valuable knowledge while having a good laugh.

Key Features

- **Comprehensive Coverage:** Explore a wide array of topics, including Return on Security Investment (RoSI), cybersecurity in energy management, and the security of smart devices.
- **Practical Strategies:** Discover actionable steps to enhance your security posture, from basic hygiene practices to complex strategic implementations.
- **Psychological Insights:** Understand the human element in cybersecurity, with chapters on the security mindset, overcoming cognitive biases, and building a cyber-resilient culture.
- **Advanced Technologies:** Delve into cutting-edge topics like quantum computing, 5G security, and the latest in deception technologies.
- **Real-World Case Studies:** Learn from detailed case studies that illustrate the application of cybersecurity principles in various industries.
- **Engaging Humor:** Enjoy "Cyber Chuckles" scattered throughout the book, ensuring that even the most complex topics are accessible and enjoyable.

Designed for a diverse audience ranging from cybersecurity professionals and IT managers to business leaders and students, *Navigating the Cyber Maze: Insights and Humor on the Digital Frontier* is your ultimate guide to the digital frontier. Whether you are looking to enhance your technical skills, understand the broader impact of cybersecurity, or simply enjoy a good read, this book is your essential companion in the ever-evolving cyber landscape. Dive in and equip yourself with the knowledge and strategies to navigate the cyber maze with confidence and a smile.

Security, Audit and Leadership Series

Series Editor: Dan Swanson, Dan Swanson and Associates, Ltd.

Winnipeg, Manitoba, Canada.
The ***Security, Audit and Leadership Series*** publishes leading-edge books on critical subjects facing security and audit executives as well as business leaders. Key topics addressed include Leadership, Cybersecurity, Security Leadership, Privacy, Strategic Risk Management, Auditing IT, Audit Management and Leadership

The Shortest Hour: An Applied Approach to Boardroom Governance of Cybersecurity
Lee Parrish

Global Audit Leadership: A Practical Approach to Leading a Global Internal Audit (GIA) Function in a Constantly Changing Internal and External Landscape
Audley L. Bell

Construction Audit: Building a Solid Foundation
Denise Cicchella

Continuous Auditing with AI in the Public Sector
Lourens Erasmus and Sezer Bozkus Kahyaoglu

Ironwill 360° Leadership: Moving Forward: Unlock Twelve Emerging Trends for Forward Thinking Leaders
Douglas P. Pflug

The CISO Playbook
Andres Andreu

Leveraging Blockchain Technology: Governance, Risk, Compliance, Security, and Benevolent Use Cases
Shaun Aghili

The Closing of the Auditor's Mind?: How to Reverse the Erosion of Trust, Virtue, and Wisdom in Internal Auditing
David J. O'Regan

Radical Reporting: Writing Better Audit, Risk, Compliance, and Information Security Reports (Second Edition)
Sara I. James

Team Intelligence: A New Method Using Swarm Intelligence for Building Successful Teams
Mohammad Nozari

The Gardener of Governance: A Call to Action for Effective Internal Auditing
Rainer Lenz and Barrie Enslin

Navigating the Cyber Maze: Insights and Humor on the Digital Frontier
Matthias Muhlert

Navigating the Cyber Maze
Insights and Humor on the Digital Frontier

Matthias Muhlert

CRC Press
Taylor & Francis Group
Boca Raton London New York

CRC Press is an imprint of the
Taylor & Francis Group, an **informa** business

Designed cover image: © Shutterstock

First edition published 2025
by CRC Press
2385 NW Executive Center Drive, Suite 320, Boca Raton FL 33431

and by CRC Press
4 Park Square, Milton Park, Abingdon, Oxon, OX14
4RN

CRC Press is an imprint of Taylor & Francis Group, LLC

ISBN: 978-1-032-91037-6 (hbk)
ISBN: 978-1-032-91255-4 (pbk)
ISBN: 978-1-003-56218-4 (ebk)

DOI: 10.1201/9781003562184

Typeset in Sabon
by KnowledgeWorks Global Ltd.

Contents

Forewords

When Matthias first contacted me to ask me to review an article he was preparing, I thought, "Oh No" – we don't need any more controls catalogues, ontologies, checklists, or "how to do a risk assessment" essays. I was delighted to see that it was none of these, but a really thoughtful and unique article about the journey an organization can take in maturing its approach to managing cyber risks, now and into the future. Wow, this was really different orthogonal thinking that actually made me reflect on my own maturity journey and career with a number of organizations and how we can continue to strive to improve even when we think we are at the peak and the end of the journey.

The problem was that this was a 30-page essay written in excellent business language! It wasn't suitable in its current form as an academic article as it wasn't a literature review, nor was it a scientific experiment or evidence-led review and proposal. But it was too long and deep to appear as an article in a cybersecurity magazine – so it went out on LinkedIn. "What a waste" I thought.

Fast forward a year and it forms a substantial part of Matthias' new book *Navigating the Cyber Maze: Insights and Humor on the Digital Frontier.* This is a "must have" manual for all Chief Information Security Officers (CISOs) and aspiring CISOs. It introduces new concepts to cybersecurity, like "Dynaxity" as a means of understanding and dealing with the ever-complex changing nature of the threats and the increasingly intricate digital and technology landscape we operate within. It covers difficult topics like team dynamics, the psychology of attackers and the attacked, dealing with uncertainty and "black swans", together with advanced and difficult fast-moving areas such as artificial intelligence (AI), and intractable problems like slow moving, vulnerable, operational technology – and even a brilliant adaptation of the Black-Scholes-Merton formula for Options Pricing applied to cybersecurity to help with evaluating cybersecurity investments and strategies.

WARNING – this book is not an easy read as each chapter makes you think deeply about what and how you operate. If you let it – every chapter will deepen and challenge and extend your thinking – all in a good way, but you have to put the effort in and expect to feel uncomfortable along the way. Let me tell you, it really is worth it!

Robert Coles (Non Executive Director/Board Chair/Advisor/University Governor)

When Matthias told me he was writing a book, I wasn't surprised. I'd already witnessed the birth of "Cyberfantastic" (complete with its very cool logo) and read his thought-provoking, eloquent articles. It made complete sense that a book would be the next step. What did catch me off guard, however, was his request for me to write a foreword for said book. Writing a foreword is a new one on me and was particularly nervous about doing his new pride and joy the justice it deserves.

Matthias is someone I've admired since we first met – an extremely thoughtful, funny, and above all, genuine person whom I'm proud to call a friend. How could I possibly refuse?

I have been in the Technology arena now for about 35 something years now, and I've seen numerous changes, but I've also observed just as many repeating cycles and trends. Remember the great developer shortage of the late 1980s and early 1990s and the rush to retrain creatives to fill an ever-evolving void? Sound familiar? Or when thin clients were supposed to rule the world and spell the end of power-hungry local devices. Yet here we are in a race to adopt more powerful machines that are able to run AI models locally? What about the 1980s panic that computers would replace us all (with "The Terminator" amplifying our social anxieties – hello, Skynet)? It's all sounding very familiar.

With these recurring patterns, we've also seen an evolution in threats. Or have we? Are today's threats truly so different from those we faced in the 1990s? Yes, they've increased in complexity and scale, but the fundamental challenges remain surprisingly similar. For a fascinating perspective on this, I'd recommend "The Cuckoo's Egg" by Clifford Stoll. Written in 1989, it highlights many of the same issues we grapple with today – about 35 years later. Well, except for patching. We've all solved that now ... right?

So, we have these repeating cycles, persistent problems, and the ever-present Pied Piper of silver bullets promising to solve all our issues if we just follow along for a while. Where am I going with this gentle meander down memory lane? Toward the need for fresh thinking and a focus on the human elements of cybersecurity.

This is precisely what drew me to Matthias's book, as it did to his earlier articles. He offers a fresh outlook, taking conventional wisdom and giving it just the right twists and turns. He makes complex concepts digestible and entertaining – perfect for the frenetic pace of our lives today. We can dive in, extract what we need, crack a smile, and get back to work.

What truly sets this book apart, though, is its focus on the human element. As I've matured (certainly not like a fine wine, as my wife is quick to remind me), this aspect has increasingly resonated with me. While technology and processes will inevitably change, the humans at the heart of it all remain constant. Despite spectacular advancements in AI, we still need people to make key decisions, to empathize, and yes, to feed and water our new AI overlords.

More importantly, we need to look after these humans. We often read about the human element in the context of building business relationships, communicating with board members, or conducting awareness training (all of which are still extremely important and are covered in this book). But how do we, as CISOs and security professionals, develop strategies to not just cope, but flourish in this challenging industry?

Few books have delved into this area as Matthias has done here, with six chapters dedicated to the human perspective. This focus on our well-being as security professionals is not just refreshing – it's essential.

In essence, what Matthias offers in this book is a blend of timeless wisdom and cutting-edge thinking, all delivered with a healthy dose of humor and humanity. It's a guide not just to navigating the complex world of cybersecurity, but to thriving within it.

So, whether you're a seasoned professional looking for fresh perspectives or a newcomer trying to make sense of this wild and exciting field we love, you're in for a treat!

Stephen Bennett (Group Chief Information Security Officer)

I met Matthias last year at BlackHat Saudi Arabia in person. I've known of him, and we sat down (along with my 20-year-old daughter – her age at that time) and discussed a deeper concept of cybersecurity. If I am being selfishly honest, I get bored talking about cybersecurity with most people. It's the same old, and usually I'm at these events because I am being paid to be "On" and always talking about cybersecurity, with a million questions coming up to me afterwards. It's my job. But Matthias and I discussed it at a level that it didn't feel like my job anymore. It was as if you're a bored martial arts teacher, and someone comes along and shows you a different way of looking at the art and brings back the spark. This was that, as cybersecurity and martial arts are not very different. They are both arts of self-defense, at least from our sitting views.

Matthias brings a unique blend of practical expertise and philosophical insight to the field of cybersecurity. His ability to intertwine the technical aspects of cybersecurity with mindfulness and human-centric approaches was a revelation. In our conversations, he spoke about the importance of a strategic pause in the midst of a cybersecurity crisis, akin to the moment of stillness before a martial arts master strikes. This mindfulness approach not only enhances the effectiveness of the response but also ensures that decisions are made with clarity and precision.

Chapter 22 of his book, "When the Tide Goes Out: Dealing with the Breach (with Integrated Mindfulness and Human Elements)," particularly exemplifies this mindset. Matthias advocates for a calm, methodical approach to breach management, integrating mindfulness to handle the initial shock and maintain focus during forensic investigations. He emphasizes transparent communication, mindful of the human factors at play, which helps in managing the crisis with empathy and effectiveness.

Similarly, in Chapter 10, "Deception and Misdirection in Cybersecurity," Matthias draws parallels between the psychological tactics used in martial arts and those employed in cybersecurity. He discusses the use of deception not just as a defensive measure, but as a strategic tool to mislead and outmanoeuvre attackers. This chapter highlights the importance of understanding human behavior and using it to our advantage in creating robust security frameworks.

What sets Matthias apart is his holistic view of cybersecurity as an art form that requires both technical skill and philosophical depth. His book is not just a manual on how to protect digital assets but a guide to understanding the broader implications of cybersecurity in our interconnected world. It challenges us to think beyond the immediate threats and consider the long-term impacts of our security strategies.

In essence, Matthias's work revitalizes the conversation around cybersecurity, making it not just about defense mechanisms but about fostering a resilient, mindful approach to protecting our digital and physical worlds. His insights and methodologies inspire us to look at cybersecurity through a new lens, one that values the human element as much as the technical one.

I am honored to contribute to the foreword of this remarkable book. Matthias Muhlert's *Navigating the Cyber Maze: Insights and Humor on the Digital Frontier* is a must-read for anyone looking to deepen their understanding of cybersecurity and embrace a more holistic, mindful approach to this ever-evolving field.

Lance James (CEO @ Unit 221B, Hacker, Coach & Mentor, Board & Security Industry)

Before I found this book, I kept looking for it everywhere. Matthias is a practitioner and a communicator, and I learned quite a bit reading his work.

Gadi Evron (Building a world-class AI security company (CEO at Knostic | CISO-in-Residence)

Cybersecurity is a fascinating field of work; in today's digital age, where every click, tap, and swipe can potentially open a door to cyber threats, cybersecurity has never been more crucial. A close-knit community of cyber defenders fights new attacks every day. This constant need to adapt to the evolution of cyber threats ensures that the defenders' work never gets stale. But it can also make cybersecurity overwhelming for novices and difficult for old hands to keep up to date in all fields. This is where *Navigating the Cyber Maze: Insights and Humor on the Digital Frontier* steps in, transforming the labyrinth of cybersecurity into an engaging journey that is both enlightening and entertaining.

The author has masterfully blended the essential knowledge needed to navigate this ever-changing terrain but does so with a refreshing twist – humor. From tales of notorious hacks to practical tips for safeguarding your digital footprint, each chapter is crafted to educate and amuse. What sets this book apart is its ability to demystify cybersecurity without diluting its importance. This unique approach not only makes the learning process more enjoyable but also ensures that the key messages resonate long after the last page is turned.

In *Navigating the Cyber Maze: Insights and Humor on the Digital Frontier*, you will find a roadmap through the intricate world of cybersecurity. In the opening chapters 0–2, the author sets the stage by highlighting the pervasive impact of cybersecurity. This deep penetration into all areas of organization and life requires that we no longer see cybersecurity as a cost center but as what it is: a driver of value. The challenge here is that the value of the most successful cyber programs is particularly difficult to determine, precisely because no successful attack has taken place. The book provides concrete recommendations for cyber professionals and stakeholders on how to determine and optimize the value generated in a comprehensible manner.

The book goes beyond traditional technical discussions, adding the psychological, cultural, and probabilistic dimensions of cybersecurity from a defender's and attacker's point of view in the following chapters 3–10. These insights are particularly relevant in today's environment, where human factors often play a pivotal role in security breaches and an overwhelming number of threats and alerts require prioritization under uncertainty.

Matthias Muhlert then draws on his own decades of experience and best practices from his global network to provide actionable recommendations for building a sustainable cyber organization in Chapters 11–15. The daily battle against attacks can feel like tilting at windmills if it is not well managed. It is not without reason that burnout is a topic that cannot be ignored in the cyber community. It is therefore highly desirable that not every organization reinvents the wheel, but that experiences are shared. The book makes a decisive contribution to this.

One of the standout features of this book is its exploration of advanced concepts and emerging technologies in Chapters 16–21. As the founder of a cybersecurity startup, I've seen firsthand how quickly the landscape changes, making it crucial for everyone – from cyber experts to business executives – to stay informed and prepared. Whether AI, Internet of Things or Quantum Technology, there is hardly a technological megatrend that does not influence cybersecurity – and is influenced by it at the same time. We cannot duck away from it. As with every technological leap, the exciting question is how we can ensure that the advantages outweigh the disadvantages. Centuries ago, the invention of printing made mass education possible – but also disinformation. Today, the same applies to AI and its ilk. It is up to the defenders to seize the opportunities and ward off the dangers. Cyber practitioners and solution providers are in the same boat: With the right entrepreneurial mindset, we can succeed. The book provides decisive impulses for this.

The author concludes by providing a mental and organizational compass to successfully navigate the cyber maze, to learn from successes and failures and to grow from the task – as an organization, but also as an individual.

But see for yourself. Whether you are a seasoned professional, a curious novice, or someone simply looking to enhance your cyber literacy: As you embark on this journey through the digital frontier, may you find clarity, confidence, and perhaps even a few laughs along the way. The cyber maze may be complex, but with the right guidance, it becomes a navigable – and enjoyable – adventure.

Welcome to the world of cybersecurity, seen through a lens that is as sharp as it is witty. Enjoy the journey.

Dr. Tobias Lieberum (Founder & Managing Director CyberDesk GmbH)

The digital landscape is a complex and ever-evolving terrain, fraught with hidden perils and treacherous paths. *Navigating the Cyber Maze: Insights and Humor on the Digital Frontier* is more than just a guide; it is a comprehensive expedition through this intricate territory. This book stands out not for its unique blend of technical expertise, managerial acumen, and a touch of human insight.

By covering wide range of concepts from cybersecurity, psychology, economics, and management, this book offers a truly multidisciplinary perspective. This approach is essential in today's world, where cyber threats are increasingly sophisticated and require multifaceted solutions.

Navigating the Cyber Maze: Insights and Humor on the Digital Frontier provides an interesting cybersecurity overview for both seasoned security professionals and those new to the field. The book's emphasis on building a cyber-resilient culture and understanding the human element in cybersecurity is particularly essential.

In an age where cyber threats materialize every single day, this book offers support to navigate the complexities of the digital frontier. What is important, it does that with a great sense of humor – which is Matthias' hallmark.

Dr. Joanna Świątkowska (Deputy Secretary-General, European Cyber Security Organisation (ECSO))

Preface: A Guide to Navigating This Book Like a Cybersecurity Pro

Welcome to your cybersecurity journey! If you're looking for a blend of humor, practical advice, and deep insights into the ever-evolving world of cybersecurity, you've picked up the right book. As you embark on this adventure, you'll not just navigate through complex concepts and emerging threats; you'll do so with a smile, equipped with strategies to tackle whatever digital curveballs come your way.

This book is designed not just to inform you, but to transform your approach to cybersecurity. By the end, you'll have gained:

- **A Deeper Understanding:** You'll explore key concepts and the latest trends in cybersecurity, getting to grips with how these can impact you and your organization.
- **Practical Strategies:** Beyond theories, you'll discover practical steps to enhance your security posture, from basic hygiene to complex strategies to shield your digital assets.
- **Enhanced Problem-Solving Skills:** Through the use of frameworks like the Cynefin Framework and principles like Occam's Razor, you'll sharpen your ability to tackle cybersecurity challenges with strategic thinking.
- **A Sense of Humor:** Yes, cybersecurity is serious, but who says it can't be fun? You'll find "Cyber Chuckles" scattered throughout to lighten the mood and make complex ideas more accessible.

AND NOW: ENTER THE BOOK LOOP

```
// Begin Book Loop
WHILE (book not finished)
{

   PRINT "Start with Cyber Chuckles";
   // Check if Cyber Chuckles piqued your interest
   IF (interest in Cyber Chuckles == TRUE)
   {
       PRINT "Dive into the chapter from the beginning";
   }
   // Revisit Cyber Chuckles to reinforce the concepts
   PRINT "Finish with Cyber Chuckles again to see if it sticks";
   // Engage with the book chapters in random order

   PRINT "Select any chapter next; they are designed to stand alone";

}
// Upon completing the book, proceed with two tasks
```

```
PERFORM {
   1. Tell your friends about this great book;
   2. Continue your reading and learning journey;
      - Read foundational texts like "How to Measure Anything in
Cybersecurity Risk",
        works by Bruce Schneier, and "Cyber Defense Matrix: The
Essential Guide to Navigating the Cybersecurity Landscape".
      - Explore outside the cyber realm with books like "Thinking, Fast
and Slow" by Daniel Kahneman,
        or "Antifragile: Things That Gain from Disorder" by Nassim
Nicholas Taleb.
}
// Function to navigate chapters with attributes
FUNCTION NavigateChapter(title, focus, attributes) {
   PRINT "Chapter Title: " + title;
   PRINT "Focus Area: " + focus;
   PRINT "Attributes: " + attributes;
}
// Introductory and Foundational Concepts
NavigateChapter(
   "Chapter 1: The ROI Rollercoaster: Demystifying Return on Security
Investment (RoSI)",
   "Introductory and Foundational Concepts",
   ["Understanding RoSI", "Fundamentals", "Challenges"]
);
NavigateChapter(
   "Chapter 2: Management Mantras or Mayhem? Understanding Possible Risk
Statements from Management",
   "Introductory and Foundational Concepts",
   ["Decoding Management Speak", "Translating Business Risks", "ERM
Alignment"]
);
// Mindset and Cultural Aspects
NavigateChapter(
   "Chapter 3: Leading Like a Hacker: Security Mindset for Everyone",
   "Mindset and Cultural Aspects",
   ["Balanced Ownership", "Universal Leadership Principles", "Team
Ecosystem"]
);
NavigateChapter(
   "Chapter 4: The Security Mindset Paradox",
   "Mindset and Cultural Aspects",
   ["Impact and Cost", "Common Cognitive Biases", "Strategies to Overcome
Biases"]
);
// Dealing with Uncertainty and Complexity
NavigateChapter(
   "Chapter 7: When the Fog Rolls In: Dealing with Uncertainty in Cyber
Security",
   "Dealing with Uncertainty and Complexity",
   ["Cognitive Reappraisal", "Strategic Ignoring", "Stochastic Modeling"]
);
NavigateChapter(
   "Chapter 8: Dynaxity and Beyond: A Model for the Modern Cyber
Landscape",
```

```
   "Dealing with Uncertainty and Complexity",
   ["Dynaxity Explained", "Policy Framework", "Hypothetical Scenarios"]);
// Advanced Topics and Technologies
NavigateChapter(
   "Chapter 10: Deception and Misdirection in Cybersecurity",
   "Advanced Topics and Technologies",
   ["Shifting Focus", "Evolving Tactics", "Future of Deception"]
);
NavigateChapter(
   "Chapter 16: Quantum Leap? Leverage Quantum for Security and
Efficiency",
   "Advanced Topics and Technologies",
   ["Quantum Internet", "Quantum Blockchain", "Energy Optimization"]
);
// Specialized Areas and Practical Implementations
NavigateChapter(
   "Chapter 18: Operational Technology (OT) Security",    "Specialized
Areas and Practical Implementations",
   ["High-Level Overview", "Detailed Implementation", "Real-World
Examples"]
);
NavigateChapter(
   "Chapter 19: Weaving the Security Fabric",
   "Specialized Areas and Practical Implementations",
   ["Zero-Trust", "Emerging Technologies", "Edge Computing"]
);NavigateChapter(
   "Chapter 20: Rethinking Access for AI",
   "Specialized Areas and Practical Implementations",
   ["Evolution of Access Control", "Dynamic Access Control", "Security
Implications"]
);
// Holistic and Forward-Looking Perspectives
NavigateChapter(
   "Chapter 22: Navigating Cybersecurity: The Integrated Cybersecurity
Compass (ICC)",    "Holistic and Forward-Looking Perspectives",
   ["Navigational Tools", "Economic and Behavioral Considerations", "Core
Security Principles"]
);
NavigateChapter(
   "The Last Chapter: Embracing a Principle-based Philosophy in the
Cybersecurity Odyssey",
   "Holistic and Forward-Looking Perspectives",
   ["Principle-Based Navigation", "Empirical Foundations", "Philosophical
Approaches"]
);
```

PRINT "Have fun and advance in your cybersecurity career!";

This preface, structured like a pseudo-code, outlines how to effectively utilize the modular design of this book. Each chapter is independent, allowing you to explore topics in any order that suits your interest or need. The use of "Cyber Chuckles" at the beginning and end of each chapter is designed to engage your interest and solidify your understanding with a touch of humor.

By treating this book as an interactive loop, you are encouraged to continuously reflect on what you've learned and apply these insights in various contexts – just like iterating through a loop in programming until the desired condition is met.

OVERALL STRUCTURE

- **Introductory and Foundational Concepts:** The initial chapters introduce fundamental concepts and set the stage for more detailed discussions.
- **Mindset and Cultural Aspects:** Chapters focusing on the security mindset, psychological principles, and building a cyber-resilient culture follow, emphasizing the importance of human factors and organizational culture in cybersecurity.
- **Dealing with Uncertainty and Complexity:** These chapters delve into handling uncertainties and complex scenarios in cybersecurity, providing frameworks and models for better decision-making.
- **Advanced Topics and Technologies:** The middle chapters cover advanced security techniques, financial models, deception strategies, and emerging technologies like quantum computing and 5G.
- **Specialized Areas and Practical Implementations:** The later chapters focus on practical approaches to specific areas like operational technology, artificial intelligence (AI), and access control in the context of cybersecurity.
- **Holistic and Forward-Looking Perspectives:** The final chapters explore comprehensive strategies like the Integrated Cybersecurity Compass (ICC) and philosophical approaches to cybersecurity, wrapping up the book with a forward-looking perspective.

SEQUENCE OF CHAPTERS

Introductory and Foundational Concepts

- Preface: A Guide to Navigating This Book Like a Cybersecurity Pro
- Chapter 0: Because this is how IT guys count
- How Cybersecurity Protects Our Daily Lives Beyond Computers and Phones
- De La Soul: Me, MYSELF, and I
- Me, Myself, and I: The Cybersecurity Groove
- Chapter 1: The ROI Rollercoaster: Demystifying Return on Security Investment (RoSI)
- Chapter 2: Management Mantras or Mayhem? Understanding Possible Risk Statements from Management

Mindset and Cultural Aspects

- Chapter 3: Leading Like a Hacker: Security Mindset for Everyone
- Chapter 4: The Security Mindset Paradox
- Chapter 5: Psychological Principles to Boost Cybersecurity Awareness
- Chapter 6: Building a Cyber-Resilient Culture

Dealing with Uncertainty and Complexity

- Chapter 7: When the Fog Rolls In: Dealing with Uncertainty in Cyber Security
- Chapter 8: Dynaxity and Beyond: A Model for the Modern Cyber Landscape
- Chapter 9: Black Swans, Grey Rhinos, and Your Bottom Line: Adapting Financial Models to Cyber Risk

Advanced Topics and Technologies

- Chapter 10: Deception and Misdirection in Cybersecurity
- Chapter 11: Red Teaming, Blue Teaming, Purple Teaming, and Yellow Teaming: A Holistic Approach to Security Assessments

- Chapter 12: Is Cyber Fantastic Still Fantastic? Re-evaluating the Future of Cyber Security
- Chapter 13: Unraveling Complexity – The Silent Adversary in Cybersecurity
- Chapter 14: Navigating Cybersecurity with Precision
- Chapter 15: Adapting Kanban for Cybersecurity Excellence
- Chapter 16: Quantum Leap? Leverage Quantum for Security and Efficiency
- Chapter 17: 5G and Beyond – Securing the Future of Connectivity

Specialized Areas and Practical Implementations

- Chapter 18: Operational Technology (OT) Security
- Chapter 19: The Beautiful World of AI
- Chapter 20: Rethinking Access for AI

Holistic and Forward-Looking Perspectives

- Chapter 21: When the Tide Goes Out: Dealing with the Breach (with Integrated Mindfulness and Human Elements)
- Chapter 22: Navigating Cybersecurity: The Integrated Cybersecurity Compass (ICC)
- The Last Chapter: Embracing a Principle-based Philosophy in the Cybersecurity Odyssey

Now that you are familiar with the structure, flip the page and begin your journey through the cybersecurity landscape, armed with knowledge and ready to tackle digital challenges with confidence and a smile.

Cyber Chuckles: A Taste of What's to Come

- **Why did the cybersecurity expert bring a ladder to work?**
  - Because they wanted to reach the top level of security!
- **Knock, knock**
  - Who's there?
  - Anomaly detection.
  - Anomaly detection who?
  - Just checking if everything's normal around here.
- **How many OT security professionals does it take to change a light bulb?**
  - None. They just need to ensure the system is resilient enough to handle it automatically.

Enjoy your reading journey and remember: cybersecurity doesn't have to be all serious – it can be engaging and fun too!

About the Author

Matthias Muhlert exemplifies Information Security leadership with over two decades of transformative contributions. His career is marked by a commitment to empowering teams, optimizing processes, and leveraging cutting-edge technology to ensure operational excellence and strategic alignment with business goals. Currently, as the "Cyber Chef for Pies, Pints, Pastries, Parties, and Pizza" at Dr. August Oetker KG and serving as ECSO CISO Ambassador for Germany, Matthias is dedicated to fortifying digital landscapes against evolving threats.

His journey includes leading global security initiatives, fostering resilient and agile security frameworks, and building robust relationships across organizational levels. Matthias's expertise is validated by certifications such as ISO 27001 ISMS Manager, CISM, CISSP, and Certified Ethical Hacker. His roles have ranged from automotive CISO to spearheading IT security in banking, showcasing his ability to navigate the complexities of cybersecurity in diverse settings.

At Oetker-Group, Matthias is responsible for developing and setting security standards across all companies, orchestrating a group-wide security community, and devising comprehensive strategies for information and OT security. His tenure at HARIBO GmbH & Co. KG involved heading the information security management system, establishing a dynamic ISMS, and pioneering an AI decision model in collaboration with other companies. As CISO at Schaeffler Technologies AG & Co. KG, he led local and global teams, introduced an information risk management methodology, and contributed significantly to international security standards.

Matthias's earlier roles include leadership positions at HELLA Corporate Center GmbH (CISO), UniCredit Bank AG (VP Information Security), and Kabel Deutschland GmbH (Main Information Security Manager), where he developed and implemented robust IT and data management strategies, led significant security development projects, and facilitated compliance with international standards. His career also spans roles as a Senior IT Security Consultant and CERT Manager at Verizon Business EMEA and as the Head of Penetration Testing Department at NESEC Gesellschaft für angewandte Netzwerksicherheit mbH.

A strong communicator with excellent interpersonal and negotiation skills, Matthias excels in identifying and mitigating security risks, conducting business process reviews, and developing appropriate solutions. His academic background includes a Bachelor of Science in Network Computing from TU Bergakademie Freiberg, and he is fluent in both German and English. Matthias is not just a strategist but a connector, bridging gaps between technology and business, ensuring security is not just a protocol but a business enabler. His efforts to craft a cohesive European cybersecurity community underscore his commitment to collaborative growth and collective security resilience.

Acknowledgements

Me, Myself, and I: The Cybersecurity Groove

Introduction

Welcome to a special chapter where we drop the beats and rhymes of gratitude and introduce the maestro behind this cybersecurity symphony. Inspired by the legendary De La Soul, this chapter is all about the people who made this book possible and a look into the life of the author. Let's get into the groove!

THE CYBERSECURITY ANTHEM

Verse I: The acknowledgments flow and foreword contributors

First off, big shout-out to **Robert Coles**,
(Non-executive Director/Board Chair/Advisor/University Governor),
Dropping wisdom like a boardroom sage, your guidance set the stage.
Stephen Bennett, the Group CISO of Dominos,
Your insights brought the beef, no need for sorrow or woes.

Lance James, CEO of Unit 221B,
Your hacker's heart and mentor's mind helped shape this spree.
Dr. **Joanna Świątkowska,** ECSO's guiding light,
Your vision keeps us steady, through every cyber fight.

Gadi Evron, AI security, world-class and high,
Your foresight sets the bar; thanks for the foreword fly.
And **Dr. Tobias Lieberum**, from CyberDesk you lead,
Your expertise and insight have truly planted a seed.

Special thanks

Annika Jerosch, just 13 but already keen,
Your quantum ideas are a techie's dream.
Sounil Yu, with your D.I.E. model,
You made sure our defenses never throttle.

Reviewers

Clément HERSSENS and **Dr. Jean-Jacques Viala,**
Your sharp eyes and feedback kept the book's integrity, holla.

Publisher

Gabriella Williams, the editor extraordinaire,
In cybersecurity, privacy, and risk, your prowess beyond compare.
Radhika Gupta from the editorial team,
Your precision and dedication brought this book to a gleam.
Dan Swanson from CRC Press,
Your support was crucial, simply the best.

CHORUS: THE CYBERSECURITY ANTHEM

It's not all about Me, Myself, and I,
It's about the team, reaching for the sky.
With a crew so tight, and minds so bright,
We're securing the future, day and night.

Verse 2: The bio beat

Meet Matthias Muhlert, in the cyber world he's known,
Leading with over two decades, his skills have grown.
From Dr. Oetker's cyber chef, guarding pies and pints,
To ECSO CISO Ambassador for Germany, always on the frontlines.

A visionary with certifications galore,
ISO 27001, CISM, CISSP, and more.
With roles in automotive, banking, and beyond,
His strategic mind keeps systems strong.

A connector of tech and business, ensuring security's a witness,
Turning threats to dismissal, with strategic finesse.
Empowering teams, optimizing processes, leveraging tech – his drive never ceases,
Building resilient frameworks and fostering unity in European cyber pieces.

Outro: Keep the beat going

Shout-out to you, the reader, your journey's just beginning,
Stay curious, stay sharp, keep the cyber beats spinning.

Part I

Introductory and foundational concepts

Chapter 0

Because this is how IT guys count

Spoiler Alert!

- Forget Chapter 1, cybersecurity protects stuff beyond your phone (shocking, we know).
- Turns out keeping the lights on (literally) involves fighting hackers.
- Who knew farms needed cybersecurity? Now your carrots can be safe from digital theft.
- This chapter ends with "Why should you care?" The answer: Because you don't want your fridge held hostage by malware.

HOW CYBERSECURITY PROTECTS OUR DAILY LIVES BEYOND COMPUTERS AND PHONES

Imagine waking up to a world where smart thermostats malfunction, cars refuse to start, and local supermarkets can't process payments – all due to cyberattacks. This might seem like a plot from a science fiction novel, yet it's a potential reality in today's interconnected world. While awareness about cybersecurity in safeguarding online accounts or smartphones is common, its critical role in more mundane aspects of daily life often remains unnoticed.

Cybersecurity extends far beyond email protection or social media safety. It is an invisible yet essential guardian in areas as diverse as home energy systems and agriculture. This chapter peels back the curtain on the unseen world of cybersecurity, revealing how it underpins many modern conveniences. From the energy that powers homes to the food supply chains, cybersecurity plays a pivotal role in maintaining the everyday rhythm of life, silently protecting against unseen threats.

SECTION 1: THE CYBERSECURITY LANDSCAPE BEYOND THE OBVIOUS

When thinking about cybersecurity, the first images that might pop into mind are of hackers targeting banks or social media platforms. But the digital shield extends far beyond these familiar frontiers. Imagine the convenience of a smart home turning into a liability or the trust placed in automotive safety systems being compromised by a line of malicious code.

Automotive systems: The connected road

Today's cars are more than just engines and wheels; they are computers on wheels. With Global Positioning System (GPS) navigation, Bluetooth connectivity, and even remote start functions, modern vehicles are ripe targets for cyber threats. A hacked car could mean more than just stolen data – it could be a matter of safety on the road.

DOI: 10.1201/9781003562184-2

Smart appliances: The household hack

The humble refrigerator or coffee maker is now part of the Internet of Things, often connected to the same network as laptops and smartphones. A breach in these everyday devices could lead to an invasion of privacy or even a pathway for attackers to access more critical data.

Retail point-of-sale systems (POS): The checkout challenge

The simplicity of swiping a card at a store hides a complex web of cybersecurity measures. Retail POS systems are a goldmine for data thieves, and a breach here can lead to widespread financial fraud affecting thousands of unsuspecting shoppers.

Library systems: The knowledge keepers' dilemma

Libraries, the vaults of knowledge and community hubs, are not immune to cyber threats. With extensive databases of user information and digital collections, library systems face unique challenges in safeguarding their digital and physical realms.

Fitness trackers and health apps: The personal data gym

In the quest for health and fitness, wearable devices and apps collect sensitive personal data. This data, if not well guarded, can be misused in ways that range from privacy breaches to insurance fraud.

Gaming platforms and consoles: The playful portal

Online gaming platforms and consoles are not just for entertainment. They involve financial transactions and personal data storage, becoming targets for cyberattacks that can lead to significant financial and personal data loss.

In each of these examples, the common thread is the unassuming nature of these devices and systems. They blend seamlessly into daily life, often hiding the complexities and vital importance of cybersecurity that keeps them, and by extension, the users, safe.

SECTION 2: KEEPING THE LIGHTS ON – CYBERSECURITY IN ENERGY MANAGEMENT

The comfort of a well-lit, warm home or an air-conditioned office is something many take for granted. But behind this comfort lies a complex network of energy management systems, increasingly reliant on digital technology. These systems, essential for controlling everything from heating to electricity distribution, are potential targets for cyberattacks that can cause widespread disruptions.

A chilling reality

Take, for instance, a city's HVAC (Heating, Ventilation, and Air-Conditioning) system. A cyberattack on such a system could disrupt temperature controls, leading to uncomfortable, even dangerous, living and working conditions. In severe cases, such disruptions could even impact critical services like hospitals and data centers.

The grid under siege

The power grid is another area where cybersecurity plays a crucial role. Imagine a coordinated cyberattack on the power grid; it could plunge entire cities into darkness. The implications range from inconvenience to life-threatening situations, especially in regions with extreme weather conditions.

Smart, yet vulnerable

As homes and buildings become smarter, integrating systems like automated lighting and energy-efficient controls, they also become more vulnerable to cyber threats. A breach in these systems could lead to energy wastage or, worse, give attackers a backdoor to more sensitive personal and corporate data.

These scenarios are not mere speculations. Incidents across the globe have shown that cyberattacks on energy management systems can have far-reaching consequences. Thus, ensuring the cybersecurity of these systems is not just about maintaining comfort but about safeguarding the very infrastructure that powers modern life.

SECTION 3: FROM FARM TO FORK – SECURING OUR AGRICULTURE

Agriculture, an essential yet often overlooked component of our daily lives, has embraced technology in ways that might surprise many. From GPS-guided tractors to automated irrigation systems and data-driven crop management, modern agriculture is a field ripe for cyber threats. The security of this sector is not just about protecting data; it's about ensuring the very sustenance of communities worldwide.

The high-tech harvest

Consider precision agriculture, where farmers use sophisticated technology to monitor and manage their crops. A cyberattack on these systems could lead to incorrect data, resulting in poor crop decisions and significant agricultural losses.

Supply chain vulnerability

Beyond the farm, the agricultural supply chain is a complex network involving transport, storage, and distribution systems. A cybersecurity breach here could disrupt food supply chains, leading to shortages or spoilage, impacting both producers and consumers.

Ransomware on the ranch

A growing concern in agriculture is ransomware attacks. Cybercriminals targeting farms and agricultural businesses can lock vital data, crippling operations until a ransom is paid. Such attacks can have devastating effects, especially for small-scale farmers and local communities.

The global plate

In an interconnected world, a cyber threat in one region's agricultural sector can have ripple effects globally, affecting food prices, availability, and even international relations. The security of agricultural data and systems is, therefore, not just a local concern but a global necessity.

The integration of cybersecurity in agriculture is a critical, yet often unheralded aspect of ensuring food security and stability. As the sector continues to evolve technologically, the importance of safeguarding its digital infrastructure becomes increasingly paramount, affecting everything from the local farm to the global table.

SECTION 4: WHY SHOULD YOU CARE?

Understanding the expansive role of cybersecurity in energy management and agriculture leads to a crucial question: why should the average person care? The answer lies in the interconnected nature of our modern world, where a breach in one system can ripple across multiple facets of life.

Personal impact

Consider the vulnerability of your home's smart devices or the food supply chain you rely on. A cyberattack in these areas can have direct consequences, from compromised personal data to disruptions in food availability or pricing.

Economic implications

Cybersecurity isn't just about protecting data; it's about economic stability. A breach in agricultural data could disrupt markets, affecting prices and availability of produce. Similarly, an attack on energy management systems could lead to significant financial losses, both for businesses and consumers.

Community and national security

On a larger scale, cybersecurity is a matter of national and community security. Attacks on critical infrastructure can have far-reaching effects, impacting everything from emergency services to national defense.

Global interdependence

In an increasingly globalized world, a cybersecurity threat in one area can have international repercussions. For instance, a breach in the agricultural supply chain in one country can affect food security in another.

The role of awareness and action

Understanding these implications is the first step. The next is action. This can be as simple as being more conscious of personal data sharing, supporting businesses that prioritize cybersecurity, or advocating for stronger cybersecurity policies and practices.

In the end, cybersecurity is not just a concern for IT professionals or corporations. It's a collective responsibility, one that requires awareness, education, and proactive measures from everyone. By understanding the role, it plays in seemingly unrelated aspects of life, individuals can contribute to a more secure and resilient digital world.

CONCLUSION

In examining the varied aspects of cybersecurity, from its role in the digitalization of agriculture to its importance in energy management, a key observation emerges – cybersecurity is a fundamental component of modern society. Its relevance extends beyond specialized tech sectors to touch on various elements of everyday life.

The range of examples discussed in this chapter of a way of introduction – from smart home devices to international food supply chains – demonstrates the extensive and diverse impact of cybersecurity. These instances show how cybersecurity influences our lives in ways that are often subtle yet significant. The consequences of cybersecurity breaches can vary widely, from causing minor disruptions to posing serious threats to personal safety, economic stability, and national security.

This exploration into the often-unseen areas of cybersecurity highlights the interconnected nature of today's world. It underscores the role that everyone, regardless of their profession or expertise, has in this ecosystem. Increased awareness, informed practices, and proactive engagement in cybersecurity measures contribute to a more secure digital environment for all.

As technological advancements continue to integrate more deeply into various aspects of daily life, the significance of cybersecurity is poised to increase. Addressing the challenges and embracing the opportunities presented by cybersecurity requires collective vigilance, innovation, and collaboration. The often-overlooked aspect of cybersecurity remains critical, quietly ensuring the uninterrupted integration of our digital and real-life experiences.

CYBER CHUCKLES: KEY TAKEAWAYS WITH A TWIST

Why your carrot's digital well-being might just save the world

As we've journeyed through the digitized prairies of our modern existence, from the cyber-guarded fields of agriculture to the digital fortresses protecting our energy grids, it's clear that cybersecurity isn't just about keeping your emails safe from prying eyes. It's about ensuring that your smart fridge doesn't join a botnet and start attacking global networks instead of keeping your veggies cool.

Think about it: In a world where your car refuses to start because it's sulking over a cyberattack, and your smart thermostat has decided to move to the Bahamas, leaving you in a freezing house, the absurdity of our digital dependencies becomes apparent. It's a world where your coffee maker demands a ransom in Bitcoin before it lets you have your morning brew, and your fitness tracker blackmails you over that skipped gym session.

So, why should you care about cybersecurity? Because in the grand tapestry of our interconnected lives, even the smallest thread can unravel the whole. Your digitally secure carrot is not just a vegetable; it's a symbol of resilience in the face of cyber chaos. It stands tall (or as tall as a carrot can stand), proclaiming that we've got the upper hand against digital marauders.

As we close this chapter, remember: Cybersecurity is no longer the sole concern of IT departments and tech enthusiasts. It's the unsung hero of our daily lives, keeping the digital wild west at bay, ensuring that our modern conveniences don't turn into cyber-induced calamities. So, the next time you bite into a crisp, digitally secure carrot, take a moment to appreciate the complex web of cybersecurity that made it possible. After all, in the digital wild west, it's not just the cowboys who need protection; it's also the carrots.

Chapter 1

The ROI roller coaster

Demystifying Return on Security Investment (RoSI)

Spoiler Alert!

- This chapter figures out how much fancy security stuff is actually worth it.
- Warning: Math involved (but it's the fun kind, with pie charts!).
- Turns out measuring cybersecurity benefits is like catching a greased pig – tricky and messy.
- Learn how to impress your boss with big words like "quantitative analysis" and "theoretical examples."

INTRODUCTION

In today's digital epoch, the specter of cyber threats grows increasingly menacing, evolving with relentless unpredictability and sophistication. This emergent reality propels cybersecurity from a mere technical safeguard to a paramount strategic imperative. Central to navigating this imperative is the nuanced concept of Return on Security Investment (RoSI), a vital metric offering organizations a means to quantify the economic value derived from cybersecurity endeavors. Distinct from traditional return on investment (ROI), RoSI transcends simple profit calculations to evaluate the effectiveness of security measures by juxtaposing the costs of security implementations against the potential financial repercussions averted through effective cyber threat mitigation.

The escalating complexity of the digital landscape amplifies the significance of RoSI within contemporary cybersecurity strategies. Organizations find themselves adrift in a vast sea of evolving threats, ranging from intricate phishing schemes to elaborate ransomware attacks, necessitating a quantifiable, data-driven approach to security investments. However, translating cybersecurity benefits into quantifiable metrics presents unique challenges. These challenges stem from the valuation of intangible assets and the dynamic nature of cyber risks, underscoring the need for a comprehensive understanding and application of RoSI.

This chapter aims to demystify RoSI, exploring its foundational principles, practical applications, and broader implications for cybersecurity management. By providing a structured framework for evaluating the efficacy of security measures, RoSI empowers organizations to make informed decisions, ensuring their investments not only safeguard digital assets but also align with and bolster overarching business objectives. Through this exploration, the chapter seeks to unravel the complexities of RoSI, shedding light on how it can be effectively calculated, implemented, and leveraged to fortify cybersecurity strategies amidst the digital era's challenges.

By journey's end, readers will grasp the critical role RoSI plays in marrying cybersecurity initiatives with strategic business outcomes, navigating the intricate balance between cost, risk, and reward in the relentless battle against digital threats.

DOI: 10.1201/9781003562184-3

Understanding RoSI

The RoSI emerges as a critical concept in the arsenal of organizational cybersecurity strategies, serving as a beacon for navigating the financial implications of security efforts. This section delves into the essence of RoSI, shedding light on its significance, the mechanics of its calculation, and how it diverges from the traditional ROI paradigm, thereby offering a nuanced lens through which the financial effectiveness of security measures can be assessed.

Fundamentals of RoSI calculation

At its core, the calculation of RoSI involves a meticulous comparison between the costs incurred from implementing cybersecurity measures and the financial benefits reaped from averting potential cyber threats. This equation encapsulates direct costs, including expenditures on security technologies, personnel salaries, and training programs, alongside indirect costs that might affect productivity. Conversely, the benefits segment of the equation encompasses savings from dodged incidents – ranging from data breaches and operational downtime to reputational damages – offering a holistic view of the financial impact of cybersecurity initiatives.

Distinction from traditional ROI

RoSI's introduction marks a departure from conventional ROI metrics, primarily because it grapples with quantifying the value of negative occurrences that were successfully avoided, such as potential cyber incidents. This paradigm shift necessitates a nuanced approach that harmonizes quantitative data, such as incident response costs, with qualitative factors like customer trust and regulatory compliance adherence, enriching the traditional financial analysis with a layer of cybersecurity-specific considerations.

Challenges in quantifying security benefits

One of the paramount hurdles in the RoSI calculation landscape is the challenge of distilling the multifaceted benefits of cybersecurity measures into quantifiable terms. These benefits often manifest as intangible assets, such as the enhancement of brand reputation or the attainment of regulatory compliance, presenting difficulties in their financial expression. Moreover, the fluid and unpredictable nature of cyber threats adds another layer of complexity, demanding assumptions regarding the incidence frequency, severity, and the evolving contour of potential attacks.

This exploration of RoSI's foundational principles is designed to equip organizations with the knowledge to adeptly navigate the intricacies of cybersecurity investment. By mastering the art and science of RoSI calculation and acknowledging the nuanced challenges in quantifying security benefits, organizations can align their cybersecurity strategies more closely with their holistic business objectives, ensuring each dollar invested in security yields maximal value.

In the ensuing sections, we will dissect the components integral to the RoSI calculation, offering a deeper understanding of how organizations can meticulously evaluate the costs and benefits of their cybersecurity initiatives, thereby enhancing the strategic impact of their security postures in the relentless quest to safeguard the digital frontier.

THE COMPONENTS OF RoSI

A thorough understanding of the RoSI requires an in-depth examination of its core components. This analysis illuminates the multifaceted nature of cybersecurity investments by breaking down the various elements that contribute to the RoSI calculation. A meticulous assessment of these components not only clarifies the calculation process but also underscores the comprehensive approach needed to evaluate cybersecurity measures' financial impact effectively.

Direct and indirect costs of security measures

The calculation of RoSI begins with a detailed look at the costs involved in deploying cybersecurity measures. These costs are categorized into direct and indirect expenses, each playing a critical role in the overall investment calculation.

- **Direct Costs:** These represent the tangible, quantifiable expenses directly associated with cybersecurity initiatives. Key elements include the procurement of hardware and software solutions designed to fortify digital defenses, salaries and benefits for cybersecurity personnel, expenditures on third-party security services, and the costs related to conducting security training and awareness programs. Direct costs serve as the foundational financial input for the RoSI equation, offering a clear picture of the investment side of cybersecurity endeavors.
- **Indirect Costs:** Unlike their direct counterparts, indirect costs encapsulate the more subtle, often overlooked financial impacts of implementing stringent security measures. These include potential disruptions to employee productivity as a result of rigorous security protocols, the opportunity costs incurred from reallocating organizational resources toward cybersecurity projects, and even the cultural ramifications of enforcing strict security policies on workforce morale and collaborative dynamics. Indirect costs demand careful consideration due to their significant influence on the overall effectiveness and perception of cybersecurity investments within an organization.

Savings from avoided breaches

The benefits segment of the RoSI equation focuses on the financial savings accrued from successfully thwarting cybersecurity incidents. This analysis involves two primary facets:

- **Quantifying Avoided Losses:** At the heart of RoSI's benefits calculation lies the estimation of financial savings realized through the prevention of security breaches. This includes direct savings from circumventing costs associated with fines, legal fees, incident response efforts, and remediation activities, alongside indirect savings linked to preserving customer loyalty, trust, and an organization's reputation in the marketplace.
- **Intangible Benefits:** Beyond the measurable financial savings, RoSI also accounts for the intangible advantages of robust cybersecurity measures. These encompass improved adherence to regulatory standards, enhanced competitive positioning through demonstrated security commitment, and the cultivation of a security-conscious culture within the organization. Although challenging to quantify, these intangible benefits are integral to fostering long-term organizational resilience and maintaining a competitive edge in an increasingly security-conscious business environment.

Challenges in valuation

Accurately calculating RoSI involves navigating the intricate process of valuing both the costs and benefits associated with cybersecurity initiatives. The dynamic and often unpredictable nature of cyber threats, coupled with the inherent difficulty of predicting incident frequencies and impacts, necessitates a nuanced approach to the RoSI calculation. Organizations are tasked with balancing quantitative data with qualitative insights to derive a comprehensive understanding of their cybersecurity investments' return, ensuring a holistic evaluation that captures both the tangible and intangible aspects of security measures.

This section has delineated the essential components involved in calculating RoSI, emphasizing the necessity of a balanced approach that recognizes the tangible and intangible dimensions of cybersecurity efforts. By meticulously assessing these components, organizations can acquire crucial insights into the effectiveness of their security investments, facilitating informed strategic decision-making and the optimization of their cybersecurity posture.

In the subsequent section, we will delve into the challenges organizations face in calculating RoSI, exploring the various strategies for surmounting these obstacles and enhancing the accuracy and relevance of RoSI analyses within the context of ever-evolving cybersecurity landscapes.

CHALLENGES IN CALCULATING RoSI

Navigating the calculation of RoSI unveils a unique set of challenges, intricately tied to the nuanced nature of cybersecurity benefits and the fluid landscape of cyber threats. These challenges demand strategic approaches to overcome, ensuring the accuracy and utility of RoSI as a decision-making tool. This section explores the obstacles organizations face in calculating RoSI and provides insights into strategies for addressing these challenges.

Data scarcity and quality

One of the principal challenges in the RoSI calculation is the scarcity and reliability of data regarding cybersecurity incidents and their financial impacts. Many organizations struggle with a lack of comprehensive historical data on security breaches, especially those that have successfully avoided significant incidents. Additionally, the sensitive nature of such data often means that establishing industry-wide benchmarks for comparison can be challenging, restricting the ability to contextualize RoSI calculations within a broader industry perspective.

- **Strategies for Overcoming Data Challenges:** Addressing data scarcity involves adopting proactive data collection and sharing initiatives. Organizations can engage in industry consortia, leveraging third-party research and reports to supplement their data. Simulating potential cyber incidents and their outcomes, employing tabletop exercises, and embracing cyber range practices can also yield valuable data for more accurate RoSI calculations, providing insights into potential costs and impacts of cybersecurity threats.

Quantifying intangible benefits

Another significant hurdle is quantifying the intangible benefits of cybersecurity investments. Enhancements in reputation, customer trust, and regulatory compliance are pivotal to long-term business success but present challenges in financial quantification. These benefits are crucial for a holistic understanding of RoSI yet defy straightforward valuation due to their abstract nature.

- **Approaches to Valuation:** To navigate this challenge, organizations can employ advanced analytical techniques, such as conjoint analysis or customer lifetime value models, to glean insights into the financial value of intangible benefits. Case studies and incidents from similar organizations can serve as benchmarks, offering a framework for estimating the potential impact of cybersecurity measures on customer behavior and overall business reputation.

The dynamic nature of cyber threats

The ever-evolving threat landscape introduces another layer of complexity to RoSI calculations. The frequency, nature, and severity of cyber incidents can shift rapidly, making previous assumptions and data points quickly outdated. This dynamism requires a flexible approach to calculating RoSI, one that adapts to the changing nature of threats.

- **Adapting to Threat Dynamics:** Organizations can tackle this challenge by adopting a dynamic approach to RoSI calculation, regularly updating their models to reflect the latest threat intelligence and cybersecurity trends. Incorporating real-time data and predictive analytics can significantly enhance the accuracy of RoSI estimations, allowing organizations to stay ahead of emerging threats.

Aligning with organizational objectives

Lastly, aligning RoSI calculations with broader organizational objectives and risk tolerance levels is crucial. Security investments must not only be effective in mitigating risk but also align with the organization's strategic priorities and resource allocation frameworks, ensuring a balanced approach to cybersecurity investment.

- **Strategic Alignment:** Facilitating close collaboration between cybersecurity teams, financial analysts, and strategic planners is key to aligning RoSI calculations with organizational goals. This collaboration ensures that cybersecurity investments are optimized to achieve both security and business objectives, fostering a balanced and strategic approach to resource allocation.

Addressing the challenges in calculating RoSI requires a multifaceted strategy, integrating data-driven insights, advanced analytical techniques, and strategic alignment with organizational goals. By overcoming these obstacles, organizations can enhance the accuracy and relevance of RoSI analyses, empowering informed decision-making and strategic investment in cybersecurity measures.

The subsequent section will explore strategic implementation practices for RoSI analysis within an organization's cybersecurity strategy, highlighting the process for integrating RoSI analysis into decision-making and ongoing evaluation to optimize cybersecurity investments.

RoSI IN ACTION: STRATEGIC IMPLEMENTATION

Successfully implementing RoSI within an organization's cybersecurity framework transcends mere calculation. It involves a comprehensive strategy that integrates these financial insights into the broader cybersecurity decision-making process. This section outlines a systematic approach for embedding RoSI analysis into an organization's cybersecurity strategy, ensuring that it serves both tactical and strategic purposes.

Step 1: Establishing a Baseline

- **Initial Assessment:** The foundation of effective RoSI implementation is a thorough assessment of the existing cybersecurity posture. This involves identifying current security measures, allocated resources, past incidents, and their financial impact. Establishing this baseline provides a clear starting point for future RoSI calculations and sets the stage for identifying areas where investments can yield significant improvements.
- **Defining Metrics:** Critical to this process is the selection of specific metrics that accurately reflect the costs and benefits associated with cybersecurity investments. These metrics should align with the organization's broader security and business objectives, providing a quantifiable framework for evaluating the impact of security initiatives.

Step 2: Engaging Stakeholders

- **Cross-Functional Collaboration:** Effective RoSI implementation requires the engagement of stakeholders across various functions within the organization. This includes IT, finance, risk management, and executive leadership. By fostering a collaborative environment, organizations can ensure a holistic view of cybersecurity investments and secure buy-in from all relevant parties.
- **Educating on RoSI Principles:** Educating stakeholders about the principles of RoSI, its significance, and its application in informing cybersecurity decisions is crucial. Workshops, training sessions, and regular communications can help demystify RoSI, fostering a shared understanding and commitment to data-driven decision-making.

Step 3: Integrating RoSI into Decision-Making

- **Decision Frameworks:** Integrating RoSI analysis into cybersecurity decision-making involves developing frameworks that use RoSI as a critical factor in evaluating potential investments. This could include setting RoSI thresholds for approving projects or prioritizing initiatives based on their expected return on investment.
- **Scenario Analysis:** Applying RoSI calculations in scenario analysis allows organizations to assess the potential impact of various security investments under different threat scenarios. This strategic planning tool helps in identifying the most cost-effective measures for enhancing cybersecurity resilience.

Step 4: Continuous Evaluation and Adjustment

- **Monitoring and Reporting:** Establishing mechanisms for the ongoing monitoring of RoSI metrics is essential. Regular reviews and reporting to key stakeholders ensure that the effectiveness of cybersecurity investments is continuously assessed against evolving threats and organizational priorities.
- **Iterative Improvement:** Insights gained from RoSI analysis should inform continuous refinement and adjustment of cybersecurity strategies. This iterative approach enables organizations to optimize their security investments, ensuring they remain aligned with both the dynamic cybersecurity landscape and the organization's evolving objectives.

Step 5: Sharing Successes and Lessons Learned

- **Case Studies:** Documenting and sharing internal case studies that highlight successful applications of RoSI analysis can serve as valuable educational tools.

These stories can illustrate the practical benefits of RoSI, encourage broader adoption of data-driven security practices, and foster a culture of continuous improvement.

- **Feedback Loops:** Creating feedback mechanisms that allow for the collection and integration of insights from across the organization can enhance the relevance and accuracy of RoSI analysis. This collaborative feedback is invaluable in refining the approach to cybersecurity investment and ensuring it meets the organization's unique needs.

By following these steps, organizations can ensure that their approach to RoSI analysis is not only comprehensive and strategically integrated but also adaptive to the changing landscape of cybersecurity threats. This strategic implementation of RoSI empowers organizations to make informed, data-driven decisions, optimizing their cybersecurity investments for maximum impact.

Next, we'll explore real-world applications of RoSI analysis, presenting case studies that illustrate its impact on cybersecurity investment decisions and strategy formulation.

CASE STUDIES: RoSI AT WORK

The practical application of RoSI analysis within organizational contexts highlights its transformative impact on cybersecurity decision-making and strategic planning. This section delves into several case studies that exemplify how organizations across various industries have successfully integrated RoSI into their cybersecurity frameworks, elucidating the challenges encountered, strategies employed, and outcomes achieved.

Theoretical examples of RoSI calculation

To further illuminate the practical application of RoSI, let us explore two theoretical examples that demonstrate how organizations can calculate the RoSI. These examples provide a quantitative perspective on evaluating the impact and effectiveness of security investments, offering a structured approach to making informed decisions about future cybersecurity initiatives.

Example 1: Comprehensive RoSI calculation

1. **ROI Calculation**
   - **Formula:** ROI = (Net Financial Return/Cost of Investment) × 100.
   - **Scenario:** An investment in cybersecurity measures costing $100,000 generates a net financial return of $150,000.
   - **Outcome:** The ROI for this security investment is calculated to be 150%.
2. **Reduction in Security Incidents**
   - **Method:** Measure the decrease in the number of security incidents before and after the implementation of specific security measures.
   - **Scenario:** A reduction from 50 incidents before the investment to 10 incidents post-investment.
   - **Outcome:** This results in an 80% reduction in security incidents.
3. **Downtime Reduction**
   - **Method:** Quantify the reduction in system downtime or operational disruptions following the implementation of security measures.

- **Scenario:** Downtime is reduced from an average of 10 hours per month to 2 hours post-investment.
- **Outcome:** The organization achieves an 80% reduction in downtime.

4. **Potential Losses Avoided**
   - **Method:** Estimate the financial losses prevented through the implementation of security investments.
   - **Scenario:** Security measures prevent potential losses amounting to $500,000.
   - **Outcome:** The calculation highlights the significant financial impact of the security measures in avoiding potential losses.

Example 2: Calculation using ALE, ARO, SLE, and EALE

1. **Single Loss Expectation (SLE)**
   - **Definition:** The estimated financial loss from a single occurrence of a security incident.
   - **Scenario:** The SLE for a particular risk is estimated at $10,000.
2. **Annual Occurrence Rate (ARO)**
   - **Definition:** The expected frequency of a security incident occurring within a year.
   - **Scenario:** The ARO for the risk is 0.5, indicating an expectation of half an occurrence per year.
3. **Annual Loss Expectation (ALE)**
   - **Calculation:** $ALE = SLE \times ARO$.
   - **Scenario:** Given the SLE and ARO, the ALE for the risk is calculated to be $5,000.
4. **Expected Annual Loss Expectation (EALE)**
   - **Definition:** The total expected financial loss across all identified risks within a year.
   - **Scenario:** Summing the ALE values for all risks results in an EALE of $100,000.

These theoretical examples elucidate the nuanced calculations involved in assessing RoSI, from straightforward ROI assessments to more complex evaluations involving ALE and EALE. By integrating these quantitative analyses into their cybersecurity strategy, organizations can gain a more comprehensive understanding of the financial benefits of their security investments.

Case study 1: Financial services firm enhances security posture

Background

A leading financial services firm faced increasing threats from sophisticated phishing attacks and malware, posing significant risks to customer data and trust. The firm sought to evaluate the financial justification for investing in an advanced threat detection system.

RoSI application

- **Costs:** The proposed system involved initial costs of $2 million for software acquisition, integration, and staff training.
- **Benefits:** Projected benefits included a 40% reduction in successful phishing attacks, estimated to save $5 million in potential fraud losses, regulatory fines, and reputational damage annually.
- **RoSI Calculation:** The firm calculated RoSI as (Benefits – Costs)/Costs, resulting in a positive return, indicating a financially justifiable investment.

Outcome

Following implementation, the firm observed a significant decrease in successful attacks, aligning with projected benefits and validating the RoSI-driven decision. This outcome not only enhanced the firm's security posture but also bolstered customer confidence and compliance with financial regulations.

Case study 2: Healthcare provider addresses compliance and data breaches

Background

A healthcare provider grappling with compliance challenges and the threat of data breaches aimed to strengthen its cybersecurity infrastructure to protect patient information and adhere to stringent regulatory standards.

RoSI application

- **Costs:** The investment in new compliance and security measures was estimated at $1.5 million, covering technology upgrades and staff training.
- **Benefits:** Anticipated benefits included avoidance of up to $4 million in potential penalties for non-compliance and breaches, along with maintaining patient trust and loyalty.
- **RoSI Calculation:** Applying the RoSI formula highlighted a compelling financial argument for the investment, showcasing a return that outweighed the upfront costs.

Outcome

Post-implementation, the provider successfully mitigated the risk of breaches and ensured compliance, leading to avoided penalties and sustained patient trust. The RoSI analysis played a pivotal role in prioritizing and justifying the necessary security investments.

Case study 3: Retail chain strengthens online shopping security

Background

With the surge in online shopping, a global retail chain sought to enhance the security of its e-commerce platform to protect against increasing cyber threats and ensure a safe shopping experience for customers.

RoSI application

- **Costs:** Implementation of enhanced encryption and multifactor authentication involved costs of $750,000.
- **Benefits:** The projected benefits included preventing potential losses of $3 million from data breaches and fraud, alongside reinforcing customer trust and loyalty.
- **RoSI Calculation:** The calculated RoSI underscored the financial viability of the security upgrades, with benefits significantly outweighing the costs.

Outcome

The security enhancements led to a marked reduction in incidents of fraud and breaches, securing the online shopping environment and affirming customer trust. The RoSI-driven approach facilitated strategic investment in cybersecurity measures that delivered tangible financial and reputational benefits.

These case studies exemplify the diverse applications of RoSI analysis across sectors, demonstrating its efficacy in guiding organizations toward cybersecurity investments that not only mitigate risks but also align with financial objectives and strategic priorities. By leveraging RoSI, organizations can navigate the complex cybersecurity landscape with confidence, making informed decisions that enhance their resilience and competitive edge.

Next, we will explore emerging trends and innovations in RoSI calculation, shedding light on how advancements in technology and methodology are refining the accuracy and applicability of RoSI in the dynamic field of cybersecurity.

INNOVATIONS IN RoSI CALCULATION

The landscape of RoSI calculation is continually evolving, shaped by advancements in technology and shifts in the cybersecurity paradigm. These innovations not only refine the accuracy of RoSI analyses but also broaden their applicability, offering organizations enhanced tools to navigate the complexities of securing digital assets. This section explores the cutting-edge developments that are reshaping how organizations approach the calculation of RoSI, highlighting their implications for future cybersecurity strategies.

Artificial intelligence and machine learning

One of the most significant advancements influencing RoSI calculation is the integration of artificial intelligence (AI) and machine learning (ML) technologies. These tools offer unprecedented capabilities in processing vast datasets, identifying patterns, and predicting future security incidents based on historical data.

- **Application:** AI and ML algorithms can automate the analysis of past cybersecurity incidents and their financial impacts, providing a more dynamic and accurate basis for RoSI calculations. By predicting potential threats and their probabilities, organizations can make more informed decisions about where to allocate resources for maximum security effectiveness.

Blockchain technology

Blockchain technology introduces a new dimension to RoSI calculation by offering a transparent and immutable ledger of transactions and interactions. This feature is particularly relevant for tracking and verifying the costs and benefits associated with cybersecurity investments.

- **Application:** Implementing blockchain can enhance the integrity of RoSI calculations by securely recording investment data, incident responses, and outcomes. This immutable record provides a reliable foundation for analyzing the financial impact of security measures, ensuring that RoSI calculations are based on accurate and tamper-proof data.

Advanced analytics and big data

The proliferation of big data analytics has also impacted RoSI calculation methodologies. Advanced analytics tools enable the processing of complex datasets to uncover insights into cybersecurity threats and the effectiveness of various defense mechanisms.

- **Application:** By leveraging big data analytics, organizations can conduct comprehensive analyses of security investments and their outcomes. This includes quantifying the direct and indirect costs of security measures, as well as the nuanced benefits, such as improved operational efficiency or enhanced compliance posture. Advanced analytics facilitate a more holistic approach to RoSI calculation, incorporating a broader range of factors and outcomes.

Predictive modeling

Predictive modeling techniques have grown in importance, offering a forward-looking approach to assessing the potential benefits of cybersecurity investments. This method allows organizations to forecast the likelihood and impact of future security incidents, informing more strategic investment decisions.

- **Application:** Utilizing predictive modeling, organizations can estimate the potential financial savings from averting various cyber threats through proposed security measures. This predictive insight into the cost-benefit analysis of security investments enables a proactive rather than reactive approach to cybersecurity strategy, aligning investments with anticipated threats.

These innovations in RoSI calculation represent a significant leap forward in the strategic management of cybersecurity investments. By embracing these advanced technologies and methodologies, organizations can enhance the precision, reliability, and strategic value of their RoSI analyses. Looking ahead, the continued evolution of RoSI calculation will undoubtedly play a pivotal role in shaping the cybersecurity landscape, guiding organizations in their quest to optimize security postures in an ever-changing digital world.

The conclusion will reflect on the journey through the intricacies of RoSI, drawing together the insights gained and contemplating the future of cybersecurity investment in the light of evolving calculation methodologies.

CONCLUSION: NAVIGATING THE FUTURE WITH STRATEGIC RoSI INSIGHTS

As this exploration of RoSI reaches its culmination, it's crucial to reflect on the journey undertaken and the landscape that lies ahead. The discourse around RoSI has traversed the foundational principles, dissected its core components, navigated the inherent challenges, and delved into strategic implementation practices. Moreover, it has highlighted the real-world impact through case studies and touched upon the cutting-edge innovations shaping its future calculations. This comprehensive examination underscores RoSI's indispensable role in aligning cybersecurity initiatives with strategic business outcomes amidst the digital era's multifaceted threats.

RoSI emerges not merely as a metric but as a strategic compass guiding organizations through the complexities of cybersecurity investment. It transcends traditional financial analyses by embedding the nuanced dynamics of digital threats and the multifarious benefits of cybersecurity measures into the calculus of organizational decision-making. The evolution of RoSI calculation methodologies, marked by advances in AI, blockchain, big data analytics, and predictive modeling, heralds a new epoch where cybersecurity investment decisions are informed by data-driven insights, predictive intelligence, and immutable records of efficacy.

The future of cybersecurity investment, illuminated by RoSI, beckons organizations to adopt a more nuanced, strategic approach. This entails not just a quantitative assessment of costs and benefits but also a qualitative understanding of how cybersecurity measures contribute to the overarching resilience and strategic objectives of an organization. The integration of innovative technologies and methodologies in RoSI calculations promises a more dynamic, accurate, and foresighted perspective, enabling organizations to anticipate threats, quantify potential impacts, and invest in defenses with the highest strategic returns.

In conclusion, the strategic implementation of RoSI within an organization's cybersecurity framework is an ongoing odyssey, one that requires continuous adaptation, learning, and strategic foresight. As digital threats evolve and new technologies emerge, so too must our approaches to calculating and leveraging RoSI. By embracing a principle-based philosophy underpinned by empirical evidence, progress-oriented discussions, and simplicity, organizations can navigate the cybersecurity landscape with greater confidence and strategic acumen.

Let this chapter serve as both a guide and a catalyst for organizations seeking to optimize their cybersecurity investments in the relentless battle against digital threats. In the ever-changing realm of cybersecurity, the strategic insights afforded by RoSI stand as a beacon, guiding decisions that not only protect digital assets but also propel organizations toward their strategic horizons.

As we look to the future, let us carry forward the insights gleaned from this exploration of RoSI, applying them with strategic intent and visionary purpose. In doing so, we fortify our defenses, enrich our understanding, and ensure that our journey through the digital age is navigated with wisdom, resilience, and an unwavering commitment to security.

CYBER CHUCKLES: KEY TAKEAWAYS WITH A TWIST FOR "THE ROI ROLLER COASTER: DEMYSTIFYING RETURN ON SECURITY INVESTMENT (RoSI)"

In the world of cybersecurity, where the only constants are change and the occasional urge to run screaming from the room, we've tackled the roller coaster that is RoSI. Armed with pie charts (because, honestly, everything is better with pie) and a dictionary of terms that sound impressive at corporate gatherings, we've journeyed through the mystical land of quantifying the unquantifiable.

The tale of the greased pig

Measuring the true value of cybersecurity investments is akin to catching a greased pig at a county fair. It's slippery, unpredictable, and you're going to get a bit messy. But, armed with the right strategy (and maybe a good pair of running shoes), it's not only possible but can also be somewhat entertaining. Imagine presenting to your boss that, thanks to your cybersecurity initiatives, you've successfully caught the proverbial pig, ensuring the company's digital assets are as secure as the pig in its pen.

Math: The fun kind

Who said math couldn't be fun? When it comes to calculating RoSI, we're not just crunching numbers; we're embarking on a thrilling adventure where each calculation is a step closer to proving that spending on cybersecurity isn't just a necessary evil – it's a savvy business move. Yes, we've got formulas, and yes, there are variables that seem to have been conjured up in a wizard's cauldron, but at the end of the day, it's about showing that every dollar spent wards off digital doom and gloom.

Impress with big words

Let's not forget the real reason we dive into the depths of RoSI – to dazzle and impress with our extensive vocabulary. "Quantitative analysis" isn't just a fancy way to say "doing the math"; it's a secret handshake that gets you into the club of those who know. And "theoretical examples"? They're not just hypotheticals; they're tales of what could be told around the digital campfire to inspire and educate.

So, why the roller coaster?

Because cybersecurity, much like a roller coaster, is all about highs, lows, and the occasional loop-the-loop. It's thrilling, a bit scary, but ultimately, it's about ensuring that when the ride ends, you're safely back at the platform, ready to go again. Your investment in cybersecurity is the safety harness keeping you secure as you navigate the digital wilds.

In conclusion

As you close this chapter (and maybe breathe a sigh of relief that the math is over), remember that the ROI roller coaster isn't just a ride; it's a journey. And with your newfound knowledge (and maybe a pie chart or two), you're well equipped to prove that investing in cybersecurity isn't just smart – it's essential. So, grab your calculator, your pie charts, and maybe a slice of actual pie, because who doesn't love pie? Here's to catching that greased pig and riding the roller coaster with confidence, secure in the knowledge that your cybersecurity investments are indeed paying off.

Chapter 2

Management mantras or mayhem?

Spoiler Alert!

- Decoding management-speak: "Cybersecurity threat" = "We might get hacked."
- This chapter teaches you to translate technobabble into solutions your boss can understand.
- It's like building a bridge between IT and the land of fancy suits and briefcases.
- Bonus points for learning all ten steps of the risk tolerance statement creation process (because who doesn't love a good checklist?)

MANAGEMENT MANTRAS OR MAYHEM? UNDERSTANDING POSSIBLE RISK STATEMENTS FROM MANAGEMENT

In the convoluted nexus of cybersecurity and corporate governance, deciphering the management's vernacular on cyber threats becomes pivotal. This chapter embarks on an expedition to demystify management speak, translating it into actionable cybersecurity strategies while fostering a symbiotic relationship between IT departments and the C-suite. Through this journey, we aim to formalize "Risk Tolerances" and "Risk Appetite" statements, essential for nuanced cyber risk governance and informed decision-making.

Decoding management speak about cybersecurity threats

Cybersecurity threats often get lost in translation between technical teams and senior management. The essence of decoding this language lies in understanding the underlying business risks and articulating them in a manner that resonates with the strategic goals of the organization. It's about framing cybersecurity not just as a technical challenge but as a critical business continuity and reputation issue.

Translating business risks into technical solutions

The chasm between recognizing a business risk and implementing a technical solution is vast and fraught with misunderstandings. Bridging this gap requires a concerted effort to align the company's main business objectives with its cybersecurity posture. Whether it's expanding the customer base or safeguarding the company's reputation, each objective should be meticulously mapped to specific security measures.

Building bridges between IT and the C-suite

Establishing a dialogue between IT and executive management is fundamental to achieving a cohesive cybersecurity strategy. This involves engaging key stakeholders – from senior management to the legal department – in risk tolerance discussions. By elucidating the

DOI: 10.1201/9781003562184-4

financial, regulatory, and reputational implications of cyber risks, IT can secure the necessary buy-in for implementing robust security measures.

ERM alignment process

The journey to formalizing "Risk Tolerances" and "Risk Appetite" begins with a collaborative effort between the Board of Directors, executive committees, and the InfoSec department. This alignment is crucial for translating high-level risk appetite into specific, actionable strategies within the realm of information security.

Engage key stakeholders

Key stakeholders, including senior management, the cybersecurity team, and the legal department, play a pivotal role in shaping the organization's risk tolerance. Their collective insights and priorities inform the development of a coherent risk management strategy that aligns with business objectives and regulatory requirements.

Understand business objectives

Identifying and understanding the company's primary business objectives are crucial. This understanding lays the groundwork for tailoring cybersecurity strategies that bolster the company's mission without imposing undue restrictions on innovation and growth.

Regulatory requirements

Compliance with data protection regulations and industry standards is non-negotiable. The legal department's insights into these requirements inform the risk management process, ensuring that cybersecurity measures meet or exceed regulatory expectations.

Risk assessment

A comprehensive risk assessment, conducted by the cybersecurity team, identifies potential vulnerabilities and threats. This assessment is the foundation upon which risk scenarios are built and analyzed.

Quantitative and qualitative analysis

Both qualitative and quantitative analyses are employed to evaluate the potential impacts of identified risks. This dual approach facilitates a balanced understanding of the risks' financial, operational, and reputational dimensions, guiding the development of cost-effective mitigation strategies.

Risk tolerance discussions

Through discussions with stakeholders, the organization defines its tolerance for various risk scenarios. These discussions are critical for setting realistic expectations and allocating resources to areas of greatest need.

Risk criteria and thresholds

The establishment of risk criteria and thresholds is a decisive step in formalizing the organization's risk tolerance. These criteria serve as benchmarks for evaluating the effectiveness of implemented security measures and guide future risk management efforts.

DOCUMENT THE RISK TOLERANCE STATEMENT

The risk tolerance statement is a formal document that articulates the organization's stance on specific cybersecurity risks. This document is essential for communicating the organization's risk management strategy both internally and to external stakeholders.

Regular review and update

Cybersecurity is a dynamic field, necessitating regular reviews and updates to the organization's risk tolerance statement. This iterative process ensures that the company's risk management strategy remains responsive to evolving threats and business objectives.

Development of risk appetite steps

Creating a Risk Appetite Statement involves defining the scope, authority, and accountability for managing cybersecurity risks. This process culminates in the allocation of resources to security measures designed to achieve the defined risk tolerance level while ensuring compliance with relevant regulations and standards.

Risk areas to consider for tolerance statements

A comprehensive risk management strategy addresses a wide range of risk areas, from data breaches and cybersecurity threats to compliance, financial, and reputational risks. Tailoring risk tolerance statements to these areas ensures a holistic approach to cybersecurity, safeguarding the organization's assets, reputation, and operational continuity.

IN ORDER TO GAIN ALIGNMENTS, FOLLOW THESE STEPS WITH YOUR LOCAL ENTERPRISE RISK MANAGEMENT (ERM)

Engage Key Stakeholders Involved in the Risk Tolerance Discussions – in Order of Priority

- Senior management
- Cybersecurity team
- Product development team
- Legal department

Understand Business Objectives

- Discover the company's main business objective. For example, is it to grow its customer base?
- It will almost surely involve maintaining a reputation for secure and reliable (software) products.

Regulatory Requirements

- The legal department highlights relevant data protection regulations and industry standards, emphasizing the need to comply with those requirements.
- This information influences the risk tolerance levels, as the company aims to avoid legal and regulatory penalties.

Risk Assessment

- The cybersecurity team conducts a risk assessment, identifying potential threats (e.g., hackers, insider threats) and vulnerabilities (e.g., weak authentication mechanisms, unencrypted customer data) in their software applications and IT infrastructure as is applicable within the scope assigned from the organization.

Risk Scenarios

- The risk team creates risk scenarios, such as a successful data breach resulting in customer data exposure or the theft of proprietary source code.

Quantitative and Qualitative Analysis

- The risk and security teams use both qualitative and quantitative methods to evaluate the potential impact of a data breach.
- They assess the potential financial cost of a breach (e.g., regulatory fines, legal liabilities, customer compensation), as well as the potential damage to the company's reputation and customer trust.
- Cost-Benefit Analysis: The cybersecurity team evaluates the cost of implementing additional security measures (e.g., encryption, intrusion detection systems, employee training) to reduce the risk of a data breach versus the potential financial and reputational losses from a breach.
- When selecting a framework or methodology, consider factors such as the organization's size, risk tolerance, available resources, and the complexity of the cyber risk landscape. No single framework fits all organizations, so a tailored approach that aligns with the specific needs and goals of the organization is essential.
  - **FAIR**: Factor Analysis of Information Risk (FAIR) helps organizations understand, analyze, and measure the information risk, based on factors that contribute to risk and how each of them affects each other; is the only international standard quantitative model for information security and operational risk; provides a model for understanding, analyzing, and quantifying cyber risk and operational risk in financial terms (from Bing search).
  - **NIST Cybersecurity Framework:** While not primarily quantitative, it offers a solid foundation for integrating quantitative risk analysis into a broader risk management process.
  - **COSO ERM:** Look for alignment with enterprise risk management.
  - **ISO/IEC 27005:2018:** Has improved over the years.
  - **Octave Allegro:** Good for aligning security risk management with business objectives and emphasizes the use of quantitative risk assessment techniques
  - **Monte Carlo Simulation:** Good for running thousands of simulations to model uncertain variables and calculate probabilities and potential impact more accurately.

Risk Tolerance Discussions

- In discussions with stakeholders, the cybersecurity team presents the various risk scenarios, their potential impact, and the associated costs of mitigating the risks.
- The discussions include questions about the company's willingness to invest in additional security measures and its risk tolerance levels.

Risk Criteria and Thresholds

- After thorough discussions, the stakeholders agree that the company's risk tolerance for data breaches should be relatively low.
- They set specific risk criteria and thresholds, such as a maximum of 5% probability of a data breach occurring in any given year.

Document the Risk Tolerance Statement

- Based on the discussions and analysis, the company documents its risk tolerance statement for data breaches.
- The statement outlines that the company aims to maintain robust security measures to keep the probability of a data breach below 5% per year and outlines the measures they will implement to achieve this goal.

Regular Review and Update

- The company commits to regularly review and update its risk tolerance levels for data breaches.
- Especially when introducing new software products, encountering changes in regulations, or witnessing emerging cybersecurity threats, there should be a scheduled review of risk tolerances.

DEVELOPMENT OF RISK APPETITE STEPS

In order to complete a risk appetite statement, the following process should be completed:

Scope

- The scope describes the reach of the statement, for example, "This Risk Tolerance Statement applies to the cybersecurity risk associated with data breaches in Company XYZ."

Authority and Accountability

- The formalization of the Chief Information Security Officer (CISO) **as responsible for overseeing and reporting on** the implementation of cybersecurity controls and ensuring alignment with the defined risk tolerance levels is critical to the process of alignment. The CISO can define the security requirements for the risk tolerances described above.

The CISO will report quarterly to the senior management and the board on the company's cybersecurity posture and risk status, including progress toward achieving the risk tolerance goal, as defined.

Risk Tolerance Objective

- The declaration of an objective helps an organization to strike a balance between cybersecurity investment and risk exposure. The objective might read, "Our risk tolerance objective is to maintain a data breach risk probability below (xxxx defined threshold) while optimizing cost-efficiency."

Risk Assessment and Controls

- The cybersecurity team conduct regular risk assessments to identify vulnerabilities and potential threats to our software applications and IT infrastructure. Based on the assessment results, we will implement controls proportionate to the risk levels identified, focusing on areas with the highest potential impact on data breaches.

Risk Tolerance Level

- After a thorough risk assessment and cost-benefit analysis, the organization should set a maximum tolerance level (for example, 2% probability for data breaches) occurring in any given year. For this example, this would indicate that the organization is willing to accept a 2% chance of a data breach happening annually.

Investment in Security Measures

- To achieve the defined risk tolerance level, the organization should allocate funds to build this cybersecurity initiatives. This budget is used to deploy necessary and future-proofed solutions such as:
  - Encryption protocols
  - Strengthened authentication mechanisms
  - Implement advanced intrusion detection systems
  - Regular, optimized, and security audits
  - Continuous cybersecurity training for employees

Compliance with Regulations

- Collecting and documenting regulations needed to comply is a necessary step. It must be clearly stated that the organization is committed to complying with all applicable data protection regulations and industry standards.
- They may state something like, "We will maintain a dedicated team to monitor and ensure adherence to relevant legal requirements to reduce the risk of non-compliance-related incidents."

Regular Review and Adaptation

- The risk tolerance statement is to be subject to annual review, or more frequently if significant changes occur in the cybersecurity landscape or the company's business objectives.
- The review process will involve key stakeholders, including senior management, legal, and the cybersecurity team, to assess the effectiveness of the risk tolerance approach.

Communication and Training

- Communication of risk tolerance and training is critical in the support of this consistently throughout an organization. This risk tolerance statement must be communicated to all employees, contractors, and relevant third parties. Additionally, regular cybersecurity training and awareness programs should be conducted to ensure everyone understands their role in maintaining a secure environment.

RISK AREAS TO CONSIDER FOR TOLERANCE STATEMENTS

Here are some typical risk areas for which a company should consider creating risk tolerance statements:

Data Breach and Data Security

- The risk of unauthorized access, loss, or theft of sensitive data, including customer information, financial data, and intellectual property.

Cybersecurity Threats

- The risk of cyberattacks, such as malware, ransomware, phishing, and denial-of-service (DoS) attacks, targeting the company's IT infrastructure and systems.

Business Continuity and Disaster Recovery

- The risk of disruptions to business operations due to natural disasters, system failures, or other unforeseen events, and the ability to recover critical business functions.

Compliance and Regulatory Risks

- The risk of non-compliance with relevant laws, regulations, and industry standards that govern the company's operations, data handling, and security practices.
- Third-party risks.
- The risk associated with the use of third-party vendors, suppliers, and service providers who have access to sensitive data or play a significant role in the company's operations.

Financial Risks

- The risk of financial losses due to fraud, embezzlement, errors, or inadequate internal controls.

Reputational Risks

- The risk of damage to the company's reputation due to negative publicity, customer complaints, or public perception of the company's actions or data breaches.

Privacy Risks

- The risk of mishandling or misuse of personal information, leading to breaches of privacy regulations or loss of customer trust.

Intellectual Property (IP) Risks

- The risk of IP theft, infringement, or loss of proprietary information, which could impact the company's competitive advantage.

Supply Chain Risks

- The risk of disruptions or vulnerabilities in the supply chain that could impact the company's ability to deliver products or services.

Operational Risks

- The risk of errors, process failures, or breakdowns in internal controls that could lead to operational inefficiencies or financial losses.

Human Resources Risks

- The risk associated with employee-related issues, such as employee misconduct, internal threats, and inadequate workforce training.

Physical Security Risks

- The risk of physical threats, theft, or damage to the company's assets, facilities, or equipment.

CYBER CHUCKLES: KEY TAKEAWAYS WITH A TWIST FOR "MANAGEMENT MANTRAS OR MAYHEM?"

Welcome to the cryptic world of management speak, where "synergizing our paradigms" means "please, just make it work." This chapter is your Rosetta Stone, turning bewildering management mantras into actionable cybersecurity commandments. Let's decode, shall we?

The art of translation

Imagine cybersecurity as a foreign film. Management speaks in dubbed English, while IT provides the subtitles. Sometimes, things get lost in translation, like when "We might get hacked" becomes "The apocalypse is nigh." Our job? To ensure the subtitles match the dialogue, making sure everyone stays on the same storyline.

Building the bridge of understanding

Bridging IT and the land of fancy suits requires more than a common language; it needs a translator skilled in the art of corporate jargon and technobabble. It's about turning "implementing a multilayered security infrastructure" into "building a digital moat." Suddenly, everyone's on board with the castle defense strategy.

The checklist: A love story

Ah, the ten-step risk tolerance statement creation process. It's like a romantic comedy: predictable, yet oddly satisfying. We start with misunderstanding and confusion, journey through conflict and development, and end with harmonious agreement. And just like in the movies, there's always that one character (we're looking at you, step 7) that almost derails everything with their antics.

From mantras to mastery

Decoding management speak is not just about preventing mayhem; it's about mastering the art of turning "potential synergistic risk vectors" into "things that could go very wrong." It's a noble quest to ensure that when management says, "Let's be proactive about our cybersecurity posture," everyone knows it means, "Please, let's not be the next headline."

The checklist tango

The real most valuable player (MVP) of this chapter is the checklist. It's the dance partner you didn't know you needed, leading you through the steps of engagement, understanding, and action. It's detailed, it's organized, and if followed correctly, it culminates in a performance that leaves everyone, from the boardroom to the server room, applauding.

In conclusion: A new hope

As we close this chapter, remember: Management mantras don't have to be mayhem. With a little translation, a bridge of understanding, and a love affair with the checklist, you can turn those high-level declarations into actionable, ground-level victories. So, the next time you hear, "Let's leverage our assets for maximum impact," you'll know it means, "Make sure our digital stuff doesn't get stolen."

And there you have it, folks. Another chapter decoded, another day saved. Here's to translating, bridging, and checklist-ing our way through the digital age. May the force of clear communication be with you.

Part II

Mindset and cultural aspects

Chapter 3

Leading like a hacker

Security mindset for everyone

Spoiler Alert!

- Forget suits and ties, this chapter teaches you to lead like a hacker (think outside the box, be adaptable).
- It's all about shared responsibility and building a team that can weather any cyber storm.
- This chapter even includes a bonus story about a modern IT security company, their struggles, and their ultimate triumph (because everyone loves a good underdog story).

INTRODUCTION

Leadership, at its core, is about navigating challenges and steering an organization toward its vision. However, traditional leadership models often remain anchored in singular perspectives, neglecting the multidimensional nature of modern organizational challenges. Such models may not fully address the complexities that arise from an interconnected global economy, technology-driven transformations, or the unique dynamics of sectors like IT security.

A more holistic leadership approach is needed – one that recognizes and responds to the myriad factors influencing an organization. Leadership should be about empowering everyone within the team, balancing responsibility with empowerment, and melding emotional intelligence with logical decision-making.

The modern leader needs to be adaptable, ready to evolve with changing circumstances, and equipped to understand both the granular details of their specific industry and the broader trends shaping the world.

This hypothesis reimagines leadership by integrating principles from various disciplines and addressing the limitations of previous models. It aims to present a leadership framework that is both comprehensive and adaptable, designed not only for today's leaders but also to meet the evolving challenges of tomorrow.

BALANCED OWNERSHIP

Ownership has long been heralded as a cornerstone of effective leadership. It's about assuming responsibility, driving initiatives forward, and standing accountable for outcomes. But the concept of ownership, in many traditional leadership models, often leans toward an absolute stance: a leader should take total responsibility, irrespective of the factors at play.

However, in the intricate web of modern organizations, such an absolute perspective can lead to undue pressure, unrealistic expectations, and a skewed perception of failures and successes. True leadership is not just about claiming responsibility for every outcome, but discerning when to shoulder the blame, when to distribute it, and more importantly, when to transform it into opportunities for growth.

DOI: 10.1201/9781003562184-6

Balanced ownership is about this discernment. It's understanding that while a leader is the rudder steering the ship, there are myriad external and internal factors influencing its course. Winds can change, currents can shift, and unforeseen obstacles can emerge. Leaders must own their decisions, but also recognize, appreciate, and empower their crew's contributions, leveraging collective strengths to navigate challenges.

In this holistic perspective, balanced ownership becomes a dynamic interplay of leading and collaborating, of directing and listening, and of assuming responsibility and fostering shared accountability. It's a model that encourages leaders to adapt, to learn from both triumphs and setbacks, and to continuously evolve in their leadership journey.

UNIVERSAL LEADERSHIP PRINCIPLES

Across fields and frontiers, from boardrooms to battlefields, the essence of leadership permeates every endeavor. But while its core may remain unchanged, its expression is varied, colored by the unique nuances of each domain. To reduce leadership to the lessons of a singular sphere would be to oversimplify its complexity and diminish its potential.

Drawing lessons from diverse fields isn't about cherry-picking strategies or adopting techniques in isolation. It's about understanding the universal truths of leadership that resonate across disciplines and synthesizing them into a cohesive approach.

In the arena of sports, for instance, leadership might manifest as teamwork, resilience, and strategic play. In the world of art, it could be about vision, passion, and the courage to break conventions. The corporate sector might emphasize innovation, stakeholder engagement, and ethical considerations. Even in realms as distinct as military and education, leadership principles echo shared values – commitment, integrity, adaptability, and mentorship.

The strength of a leader doesn't lie in mastering the strategies of a specific domain but in recognizing and adapting these universal principles. By integrating insights from varied fields, leaders can cultivate a well-rounded perspective. One that's enriched by diverse experiences, adaptable to varied challenges, and robust enough to stand the test of time and turmoil.

By embracing universal leadership principles, we pave the way for a model that's not confined by the limitations of a particular field but is expansive, inclusive, and holistic, poised to guide leaders through the multifaceted challenges of the modern world.

While leadership manifests differently across various domains, certain core principles resonate universally. Here's a synthesis of these principles, drawn from a myriad of fields:

1. **Commitment (Military)**: In the face of adversity, a leader remains steadfast, demonstrating unwavering dedication to their cause and team. This commitment sets the tone for the entire organization, fostering loyalty and perseverance.
2. **Teamwork (Sports)**: Just as a cohesive team is crucial to winning a game, effective collaboration is key in any organizational endeavor. A leader nurtures this spirit, emphasizing collective strengths and shared goals.
3. **Vision (Art)**: Artists bring visions to life, breaking boundaries, and redefining norms. Similarly, leaders craft a clear and compelling vision, guiding their team toward a shared future.
4. **Innovation (Corporate Sector)**: As businesses evolve, the ability to innovate becomes paramount. Leaders encourage a culture of continuous improvement, always seeking better ways to address challenges.
5. **Adaptability (Education)**: Educators adapt to the unique needs of their students. Leaders, too, must be flexible, adjusting strategies based on evolving circumstances and diverse team dynamics.

6. **Ethical Grounding (Philanthropy):** Leaders uphold strong moral principles, ensuring that decisions are made with integrity and respect for all stakeholders involved.
7. **Strategic Thinking (Chess):** Chess masters plan multiple moves ahead, anticipating opponents' strategies. In a similar vein, leaders think strategically, foreseeing potential challenges and opportunities.
8. **Passion (Music):** Musicians pour their soul into their craft. A leader's passion fuels their drive, motivates their team, and sets the stage for achieving greatness.
9. **Mentorship (Academia):** Great professors don't just teach; they mentor. Leaders should similarly guide, nurture, and invest in the growth of their team members, ensuring a legacy of strong future leaders.
10. **Resilience (Nature):** Just as nature endures and thrives amidst adversities, leaders cultivate resilience, ensuring they and their teams bounce back stronger from setbacks.

By understanding and integrating these universal principles, leaders can craft a holistic approach that resonates across domains, ensuring they're equipped to handle diverse challenges with aplomb.

ACHIEVABLE LEADERSHIP STANDARDS

In the quest for excellence, leadership paradigms often present ideals that, while aspirational, can sometimes verge on the unattainable. The vision of a leader who is perpetually decisive, unfailingly visionary, and consistently empathetic can cast long shadows of doubt and inadequacy. Such standards, while meant to inspire, can inadvertently create a chasm between the ideal and the real, making genuine leadership growth seem elusive.

However, leadership isn't about perfection – it's about progression. Here's a reframed perspective:

1. **Embracing Imperfection:** Every leader, regardless of experience or domain, will have moments of uncertainty or error. It's in these moments that growth occurs. Recognizing and learning from mistakes is more valuable than avoiding them altogether.
2. **Setting Realistic Goals:** While ambitious targets can drive a team forward, it's essential to balance aspiration with attainability. Goals should stretch the team but not to a breaking point.
3. **Continuous Learning:** Leadership isn't a destination – it's a journey. The best leaders are perpetual students, seeking knowledge, feedback, and diverse perspectives to refine their approach.
4. **Celebrating Small Wins:** Instead of waiting for monumental achievements, leaders should celebrate incremental progress. Recognizing and valuing these small milestones fosters motivation and a positive team culture.
5. **Emotional Authenticity:** Leaders don't always need to have an unwavering facade. Expressing genuine emotions, whether it's joy, concern, or even vulnerability, can forge deeper connections within a team.
6. **Balancing Vision with Ground Reality:** While a grand vision is essential, leaders must also stay connected with the on-the-ground realities, ensuring strategies are both visionary and viable.
7. **Empowering over Micromanaging:** Leaders should trust and empower their team members, allowing them to take initiative and ownership, rather than attempting to control every detail.

8. **Acknowledging Limitations:** No leader can be omnipotent or omniscient. Recognizing one's limitations and seeking collaboration or counsel is not a sign of weakness but of wisdom.
9. **Valuing the Journey:** Leadership growth isn't just about the milestones achieved but the lessons learned, relationships forged, and the character developed along the way.

By embracing achievable standards, leadership becomes a more inclusive realm, accessible to those who are committed to growth and impact. It's not about fitting into a mold but continuously molding oneself, adapting, learning, and striving for better, one step at a time.

HARMONIZING EMOTION AND REASON

The tapestry of effective leadership is woven with threads of emotion and reason. While each brings its hue and texture, their combined pattern provides a comprehensive understanding and approach. The crux lies not in prioritizing one over the other but in the delicate act of harmonization.

1. **Emotional Resonance & Rational Clarity:** Emotion provides the resonance that connects leaders with their teams, building rapport and trust. However, it's the clarity of reason that ensures decisions are strategically sound, creating a roadmap toward shared goals.
2. **Decisions Anchored in Logic:** The backbone of any decision should be logic – objective data, proven methods, and analytical rigor. Emotion offers a lens, adding depth and perspective, but decisions must always have a solid rational foundation to ensure their effectiveness.
3. **Balanced Feedback with Precision:** While empathy ensures feedback is delivered without harming morale, reasoning ensures it's precise, actionable, and directed toward tangible improvements.
4. **Navigating Crisis with Steadfast Logic:** Emotions, especially during crises, can cloud judgment. Leaders must rely heavily on reason during these times, providing a stabilizing anchor for their teams. Emotions are acknowledged, but decisions remain rooted in fact and informed analysis.
5. **Communication: Depth & Detail:** Emotional authenticity ensures messages resonate, but it's the structured logic that provides clarity, ensuring messages are not just felt but also clearly understood.
6. **Trust Through Consistency:** Emotions can foster initial connections, but trust is solidified through consistent and logical actions. Teams trust leaders who display sound judgment time and again.
7. **Ethical Judgments with Moral Reasoning:** While emotions guide our initial understanding of right and wrong, it's the systematic moral reasoning that ensures ethical decisions are consistent and uphold organizational and societal values.
8. **Visionary Planning:** Dreams and aspirations are emotional, but bringing them to reality requires methodical planning, logical structuring, and strategic foresight.

By intertwining emotion and reason, leaders create a synergy where each complements and elevates the other. Emotion without reason can drift toward impulsiveness, while reason without emotion might seem cold or distant. But together, they forge a leadership approach that's both profoundly human and strategically sound.

TEAM ECOSYSTEM

Teams, much like ecosystems, are intricate networks of interwoven relationships, each displaying its unique characteristics. While every individual plays a role, the overall system's health hinges on understanding its distinct nature and honoring the inherent structures within.

1. **Interdependent Dynamics:** While each team member has a role, the collective success hinges on their interdependent relationships. Individual accomplishments pave the way for collective achievements, underscoring the interconnected essence of teams.
2. **Structured Hierarchies:** Nature displays hierarchies, from the food chain to specialized roles within species groups. Similarly, in teams, hierarchies, often emergent, provide order and direction. While each member is essential, the system thrives on this established structure, with roles and responsibilities clearly defined.
3. **Diversity within Frameworks:** Every team ecosystem has its unique fingerprint. While diversity in skills and experiences enriches the team, it's crucial to understand and operate within the team's specific dynamics to foster cohesion and productivity.
4. **Ecosystem Analysis:** No two team ecosystems are identical. Leaders must invest time in understanding the unique characteristics, strengths, and vulnerabilities of their team's ecosystem to guide it effectively.
5. **Adaptive Resilience:** While each ecosystem has its inherent strengths, adaptability ensures longevity. By recognizing and adjusting to internal dynamics and external pressures, teams can navigate challenges more effectively.
6. **Resource Allocation with Respect to Structure:** Efficient resource management is essential. However, understanding the team's unique ecosystem allows for a more tailored distribution of resources, from time to skills, ensuring optimal outcomes.
7. **Honoring Individual Roles within Hierarchies:** Each member, regardless of their position in the hierarchy, contributes value. Yet, it's imperative to strike a balance between individual recognition and the importance of the collective structure for harmonious functioning.
8. **Shared Objectives in Diverse Ecosystems:** While every team ecosystem is distinct, a shared vision ensures that individual efforts are aligned and contribute to overarching goals. This vision must be molded to respect and fit within the specific ecosystem's characteristics.
9. **Feedback Mechanisms Tailored to the Ecosystem:** Regular, ecosystem-specific feedback ensures alignment with goals and continuous improvement. Recognizing the unique dynamics of each team allows for feedback that's both relevant and effective.

By viewing a team through the lens of an ecosystem, leaders can delve deeper into its unique dynamics. Recognizing that every ecosystem varies in its composition, strengths, and vulnerabilities allows for a more tailored and effective leadership approach. The emphasis isn't just on the individual or the collective but on understanding and honoring the nuanced interplay between the two within their unique ecosystem context.

ADAPTABLE TEAM MODE

In the modern workplace, diverse teams present both opportunities and challenges. Effective leadership recognizes the importance of individual contributions but understands that the team's collective effort is paramount.

1. **Recognizing Team Diversity:** Teams are diverse mosaics of roles, experiences, backgrounds, and, notably, generational influences. While individuals bring their unique strengths shaped by these factors, it's the synergy of these diverse elements that drives

team success. Leaders should be sensitive to these nuances, tailoring their approach to foster collective alignment.
2. **Fluid Leadership Styles:** Teams with varied compositions require adaptive leadership approaches. Flexibility is key, aiming for a balance between guidance, collaboration, and autonomy.
3. **Tailored Communication:** Diverse teams can have differing communication preferences. It's crucial to adapt communication strategies to ensure clarity and alignment, be it through formal channels, digital means, or face-to-face interactions.
4. **Resource Allocation for Diverse Needs:** Catering to the unique requirements of diverse teams can mean providing varied tools, training methods, or resources, all geared toward maximizing collective efficiency and productivity.
5. **Training & Development for All:** Different team members might lean toward specific training formats, but a range of approaches ensures that everyone's learning needs are met, further boosting team performance.
6. **Feedback Mechanisms for Every Team Member:** While feedback is universally essential, how it's delivered can vary. Employing diverse feedback mechanisms ensures inclusivity and effectiveness.
7. **Balancing Autonomy and Oversight:** Teams or members with varying levels of experience or background might either thrive on autonomy or require more structured guidance. Leaders must discern and adjust to these dynamics.
8. **Evolving with Changing Team Dynamics:** As team compositions shift and needs evolve, adaptive leadership becomes even more vital, ensuring continuous alignment with these dynamic challenges.
9. **Understanding Generational Motivations:** Each generation is motivated differently. While Gen X might value job security and a clear career path, Millennials often prioritize work-life balance and meaningful work. Gen Z, on the other hand, may seek opportunities for rapid growth and digital innovation. Recognizing these nuances is the first step in tailored leadership.

The crux of adaptable team models lies in recognizing the individual's value within the broader team framework, ensuring that the collective objective is always in focus, and harnessing the diversity that each member brings to the table.

SHARED LEADERSHIP RESPONSIBILITY

Achieving ownership within established hierarchies

Shared leadership offers a balance by merging the concept of collective ownership with the understanding that hierarchies are foundational for organizational clarity and order.

1. **The Coexistence of Shared Leadership and Hierarchy:** Shared leadership promotes collective ownership without diminishing the significance of hierarchical structures. It underscores the value of individual input within defined frameworks.
2. **Path to Ownership**
    - **Giving Information:** Team members are provided with essential details, such as tasks, goals, or projects. This step delineates direction.
    - **Processing Information:** Members assimilate and process the given details, achieving clarity about their responsibilities through discussion, reading, or experimentation.
    - **Understanding the Role:** Team members arrive at a comprehensive understanding of their tasks and associated expectations through reflection, learning, or practice.

- **Embracing Ownership:** Having gained clarity, members actively engage with their roles, taking decisive actions and responsibility for outcomes, marked by proactive decision-making and problem-solving.

3. **Collaboration within Hierarchies:** Encouraging collaboration doesn't dissolve leadership roles. It seeks to cultivate an inclusive decision-making process, allowing for everyone's voice, while retaining clear decision-making authority in designated leaders.
4. **Dynamic Leadership Roles:** While roles in shared leadership might be fluid based on required expertise, they function seamlessly within the existing hierarchy, ensuring apt individuals lead in specific contexts.
5. **Accountability and Empowerment:** Shared leadership enhances mutual accountability. While valuing everyone's contributions, a clear chain of command remains intact for final decision-making and accountability.
6. **Reinforcing Trust within Structures:** Shared leadership accentuates trust by valuing team member contributions, with the hierarchical structure guaranteeing an organized approach to task execution.
7. **Ensuring Stability:** Distributing leadership responsibilities while acknowledging hierarchies ensures the organization remains directionally coherent and benefits from diverse inputs.

Shared leadership, approached thoughtfully, can enhance ownership, empowerment, and collaboration, especially when nested within the parameters of established hierarchies, ensuring clarity, direction, and accountability.

INTEGRATING EXTERNAL FACTORS

Employing the Dynaxity model to navigate the broader environment's influence on team dynamics

Addressing the internal workings of teams and leadership models is half the equation. The external environment, marked by its dynamic and complex nature, plays an influential role in team dynamics. The Dynaxity model serves as an essential framework to determine an organization's position in this turbulent landscape.

1. **Understanding Dynaxity:** Grasping the essence of Dynaxity helps leaders anticipate the challenges brought about by rapid change and intricate complexity. Recognizing where the organization sits within this framework aids in aligning team strategies with external realities.
2. **Organizational Culture Amidst Dynaxity:** Amidst dynamic complexities, the organizational culture stands as a beacon, guiding teams. Leaders need to ensure this culture remains adaptive, resilient, and aligned with the broader organizational mission.
3. **Market Trends and Economic Fluctuations:** In a high Dynaxity environment, markets and economies can be unpredictable. Leaders should cultivate a team mindset that's ready to pivot and adapt to these unforeseen shifts.
4. **Technological Advancements:** The pace at which technology evolves adds layers to Dynaxity. Leaders should foster an always-learning approach, ensuring teams harness relevant tools and stay ahead of the curve.
5. **Regulatory Dynamics:** In an environment marked by Dynaxity, legal landscapes can shift rapidly. Leaders need to champion agility, ensuring teams remain compliant and proactively adjust to changing regulations.

6. **Societal Values and Cultural Movements:** As societal norms and values evolve, they can bring about complexities teams must navigate. Recognizing these shifts and integrating them into team strategies ensure alignment with evolving stakeholder expectations.
7. **The Competitive Landscape in a Dynaxity Context:** Rapid changes in competitor strategies add to the Dynaxity organizations face. Leaders should instill a proactive and responsive mindset, guiding teams to stay ahead in the competitive race.
8. **Global Geopolitical Shifts:** For teams operating on a global scale, the Dynaxity brought about by geopolitical changes can be significant. A well-informed leadership approach ensures teams can navigate these intricacies effectively.
9. **Positioning with the Dynaxity Model:** Leaders should consistently evaluate where the organization stands within the Dynaxity framework. This continuous assessment helps in tailoring strategies, ensuring teams remain agile and effective in the face of dynamic complexities.

By harnessing the insights from the Dynaxity model and integrating them into leadership strategies, leaders can empower their teams to thrive, adapting and excelling amidst the challenges and opportunities presented by today's intricate business environment.

PART I: CONCLUSION

Concluding reflections on a holistic leadership paradigm

The landscape of effective leadership, once believed to be static and well-defined, has transformed. The complexities of our era demand not just leaders, but visionaries who understand the interconnected systems of the modern workplace. This comprehensive leadership blueprint we've pieced together brings forth a fresh perspective, weaving together timeless principles with contemporary insights.

At the heart of our approach is **Balanced Ownership**, a reminder that leadership isn't solely about bearing burdens, but discerning where and when to intervene. Drawing from **Universal Leadership Principles**, we've ventured beyond traditional realms, gleaning insights from varied disciplines to foster a holistic leadership mindset.

The emphasis on **Achievable Leadership Standards** and the amalgamation of emotion and reason challenges conventional leadership narratives. They underscore the importance of marrying human intuition with logical processes, and of setting attainable expectations, encouraging perpetual growth.

Recognizing teams as dynamic **Ecosystems** and promoting **Adaptable Team Models** is a testament to the ever-changing nature of workplaces, emphasizing the importance of elasticity in leadership. The **Shared Leadership Responsibility** segment promotes a cooperative approach, dismantling the notion of leadership as a lone endeavor and replacing it with a vision of collective progression.

Our dive into **Integrating External Factors**, particularly through the lens of the Dynaxity model, offers a roadmap for leaders to navigate the multifaceted challenges of today's world.

What sets this blueprint apart from the foundational source materials is its intentional integration and expansion. While the source books provided invaluable insights into leadership, teamwork, and organizational culture, our approach actively addresses criticisms and gaps. By bridging these gaps, we offer a more balanced, adaptable, and realistic leadership framework, designed for the challenges and opportunities of the present and future.

In essence, leadership, as envisioned in this paradigm, is an evolving art form, requiring continuous adaptation and growth. As dynamics shift and complexities abound, a true leader thrives by understanding, adapting, and propelling forward with both wisdom and tenacity.

PART II: ALIGNMENT WITH CHANGE MANAGEMENT

Change, though inevitable, often presents a complex puzzle for organizations. Successfully navigating this requires a combination of effective leadership and a structured approach to managing transitions. The leadership framework developed in Part I provides a foundation for addressing challenges inherent in guiding teams and setting visions. In this section, we'll explore how to intertwine the leadership principles with John P. Kotter's eight-step process for change management. By integrating these elements, we aim to equip organizations with a comprehensive strategy for managing change with clarity, foresight, and resilience.

STEPS IN KOTTER'S CHANGE MANAGEMENT THEORY

Change management is more than just a series of actions; it's a strategic approach to transition teams, departments, and organizations from a current state to a desired future state. John P. Kotter, in his insightful work, delineated an eight-step process that offers a systematic method to drive and manage change. Here's a brief introduction to each step and its inherent significance:

1. **Create Urgency:** The foundation for any change initiative is recognizing the need for change. By identifying potential threats or understanding opportunities, organizations can instill a sense of urgency, making stakeholders more receptive to the change process.
2. **Form a Powerful Coalition:** Change isn't an isolated effort. Gathering a group of influential people who believe in the necessity of change can act as a powerful force, driving momentum and gathering widespread support.
3. **Create a Vision for Change:** A clear vision serves as a north star for change initiatives. By articulating what change looks like and the benefits it brings, stakeholders can understand, relate to, and rally behind a shared goal.
4. **Communicate the Vision:** A vision, no matter how compelling, is futile if not effectively communicated. Regularly sharing and reinforcing the vision ensures that everyone understands and embraces the new direction.
5. **Remove Obstacles:** Change is often met with resistance, both logistical and human. By proactively identifying and addressing barriers, organizations ensure a smoother transition.
6. **Create Short-Term Wins:** Change is a marathon, not a sprint. By celebrating short-term victories, teams remain motivated, witnessing firsthand the fruits of their efforts and the positive impacts of the change initiative.
7. **Build on the Change:** One success doesn't denote the end. It's crucial to analyze the results, learn from them, and leverage the momentum to drive further change, ensuring long-term transformation.
8. **Anchor the Changes in Corporate Culture:** For change to be sustainable, it needs to become a part of the organizational DNA. This means integrating new behaviors, processes, and mindsets into the core culture, ensuring they endure and thrive.

Incorporating these steps with the leadership principles we've previously discussed offers a harmonious blend, ensuring not just the initiation of change but its successful implementation and longevity.

INTEGRATION OF LEADERSHIP PRINCIPLES WITH KOTTER'S STEPS

Integrating leadership principles with Kotter's change model: Addressing human and organizational dynamics

Introduction: Fusing leadership principles with Kotter's steps isn't merely about mapping processes; it demands a deep understanding of human emotions, trust-building, and the diverse ways individuals adapt to change.

1. **Create Urgency**
   - **Transparency:** Avoid prolonging the worry phase with late or unclear communication about the impending change. Neglect here fuels gossip and misconceptions.
     - **Actions:** Employ clear channels and timelines for communication; proactively address potential concerns before they evolve into rumors.
2. **Form a Powerful Coalition**
   - **Credibility:** Reacting impulsively to unexpected spikes in productivity can damage leadership's credibility. Resisting the urge to abandon the course at initial resistance reinforces trust.
     - **Actions:** Monitor productivity metrics closely; provide reasoned explanations for unforeseen productivity shifts to the team.
3. **Create a Vision for Change**
   - **Participation:** While it's not advised to delegate core responsibility, allowing stakeholders to shape change goals promotes buy-in and can be a pivotal success factor.
     - **Actions:** Facilitate workshops; engage multidisciplinary teams in co-creating the change vision, ensuring participation does not dilute responsibility.
4. **Communicate the Vision**
   - **Platform for Emotions:** As productivity and morale dip, it's crucial to let employees express their feelings rather than resorting to pep talks. There's a value in mourning past processes that worked.
     - **Actions:** Arrange town-hall meetings; allow genuine expression of feelings and ensure they feel seen and heard.
5. **Remove Obstacles**
   - **Time & Early Majority of Change Curve:** Understand that profound change doesn't happen overnight; patience is key. Cater specifically to the early majority, bridging the gap between them and the early adopters.
     - **Actions:** Offer training and resources to the early majority; communicate benefits and risks comprehensively; create a roadmap respecting the natural progression of change processes.
6. **Create Short-Term Wins**
   - **Innovators and Early Adopters:** Highlight the pioneers and early supporters of change. Their buy-in can heavily influence the next wave of adopters.
     - **Actions:** Showcase success stories of innovators and early adopters; design incentives to encourage early adoption.

7. **Build on the Change**
   - **Early Majority:** The largest segment of adopters needs evidence of success before committing. By effectively communicating with and supporting them, we ensure that change gains momentum.
     - **Actions:** Offer testimonials of early successes; provide robust support mechanisms tailored to this group.
8. **Anchor the Changes in Corporate Culture**
   - **Late Majority & Laggards:** By the time change becomes mainstream, even the most resistant members of the team start to come around. The challenge lies in easing their transition.
     - **Actions:** Regularly update organizational policies to mirror new practices; develop induction programs that emphasize the new norms.

Change isn't merely procedural; it's profoundly human. Integrating leadership principles with Kotter's framework offers a dynamic approach, attending to the nuances of transparency, credibility, emotional spaces, participation, and understanding the diverse paces at which people embrace change. This holistic approach ensures changes are not just implemented but deeply rooted and embraced.

PART II: CONCLUSION

Mastering change management through alignment

Change is an inevitable aspect of organizational dynamics. Part II delved deep into the integration of leadership principles with Kotter's change management theory. Recognizing and integrating these concepts is pivotal in understanding how they interlink and amplify each other.

Kotter's structured approach offers a roadmap, a series of steps designed to ensure that change is not only effective but also sustainable. It provides organizations with a clear direction. Yet, the true essence of this journey lies in understanding and addressing the complexities of human emotions, behavior, and expectations. This balance is where the magic happens – it's a dance between the structured and the organic, the procedural and the emotional.

The leadership principles – from transparency to participation – act as the soul of this dance. They ensure that as we traverse the pathway set by Kotter's eight steps, the individual and collective concerns and motivations of team members remain central.

Bridging the gap between early adopters and the early majority, understanding the pace of change, and ensuring clarity and credibility throughout the journey are not just tasks to be checked off a list. They are signposts of a transformative journey, one that requires vision, patience, resilience, and, above all, leadership. By embracing these principles and steps, organizations are better positioned to handle change, ensuring continuity, progress, and a harmonized work environment.

PART III: A RELATABLE STORY – MODERN IT SECURITY COMPANY

Nestled within the tech hubs of Berlin, CyberGuardTech stands as a beacon of IT security prowess. Founded in the late 2000s by two German tech visionaries, the company quickly ascended the ranks to become Europe's go-to solution for cybersecurity needs. Their innovative methodologies and adeptness in thwarting cyber threats made them the pride of the continent's digital defense sector.

But as with all thing's tech, the environment is ever evolving. New-age cyberattacks grew in complexity, and technology's relentless advance meant that what was once avant-garde soon edged toward obsolescence. Emerging competitors, with their fresh takes on cybersecurity, began to challenge CyberGuardTech's once unshakable market position. Internally, processes that had been the backbone of their success now seemed cumbersome. Discontent brewed among the ranks, with employees feeling the strain of using outmoded tools amidst an ever-demanding workload.

The company's demographic makeup added another layer to the challenge. With a blend of seasoned experts and eager newcomers, there was a rich tapestry of experience and innovation. But this also led to varied viewpoints, often leading to discord rather than synergy. It was clear: CyberGuardTech needed a transformation – a recalibration of its strategies and ethos to meet the demands of the current era.

This section charts the course of CyberGuardTech, a company with a storied past and a promising future, standing at the precipice of change. Through their journey, we'll explore the challenges they faced, the change management strategies they adopted, and the leadership dynamics that were pivotal to their resurgence.

CURRENT SCENARIO AND ISSUES

In the echoing corridors of CyberGuardTech's expansive Berlin office, there was a palpable tension. The brilliance of the firm's foundational years seemed like a distant memory, overshadowed by the looming challenges of the present.

1. **Technological Stagnation:** The cybersecurity landscape is a battleground of innovation, where offense and defense are in a perpetual dance. While CyberGuardTech had once led this dance, its current systems were becoming antiquated. Many of their tools and methodologies were designed to combat threats from a decade ago. As newer, more sophisticated cyber threats emerged, the company found itself increasingly on the back foot, often reacting to breaches rather than preemptively securing systems.
2. **Operational Inefficiencies:** The processes that had once streamlined operations now added layers of bureaucracy. Approval chains were long, leading to delays in decision-making. Redundant systems overlapped in function, consuming unnecessary resources and confusing newer staff members who struggled to discern the company's once-clear operational blueprint.
3. **Talent Discontent:** There was a growing disconnect between the company's veteran staff and the influx of fresh talent. While the former held the institutional knowledge crucial to CyberGuardTech's operations, they often found themselves resistant to newer, more agile methodologies. The newer recruits, bursting with innovative ideas and versed in modern cybersecurity techniques, felt stifled by the company's reluctance to adapt. This generational chasm led to a decrease in overall morale and an uptick in staff turnover.
4. **Market Perception:** Clients began to perceive CyberGuardTech as a legacy player, capable but not cutting-edge. The burgeoning start-ups in the field, quick to adopt the latest technologies and methodologies, started to erode CyberGuardTech's client base. Loss of significant contracts and diminishing market share were becoming all too frequent occurrences.
5. **Financial Strain:** All these issues coalesced into a concerning financial scenario. Revenue streams started dwindling, and operational costs surged due to inefficiencies and the need for damage control from cybersecurity breaches.

The consequences of inaction were severe. If these problems persisted, CyberGuardTech risked not just market irrelevance but potential insolvency. Beyond financial losses, there was a genuine threat of losing the company's most valuable asset: its talented workforce. For a firm that had once been the crown jewel of Berlin's tech sector, these challenges served as a clarion call for change. CyberGuardTech stood at a crossroads, and the path it chose would determine its future in the ever-competitive realm of cybersecurity.

THE JOURNEY OF CHANGE

In the bustling metropolis of Berlin, with its rich history and blend of traditional and modern influences, CyberSecure GmbH found itself facing challenges reminiscent of many evolving tech companies in the city's vibrant tech scene.

Recognizing the Imperative for Change: Upon realization of the growing threats and internal challenges, the leadership of CyberSecure GmbH took it upon themselves to foster change. The inception of their journey began with a commitment to a dual approach: intertwining the insights from the new leadership hypothesis with the structured steps proposed by Kotter.

1. **Establishing a Sense of Urgency:** To ignite momentum, leadership outlined clear evidence of the company's stagnation, with candid data presentations and open forums where the declining market share and client feedback became central topics. The emphasis on transparency was a conscious effort to stem any tide of rumors or misinformation.
2. **Creating the Guiding Coalition:** In an era where hierarchical decisions are becoming outdated, CyberSecure GmbH turned to shared leadership. The company assembled a diverse task force, not just from the managerial echelons but from all levels, ensuring a mix of perspectives and a true representation of the company's heartbeat.
3. **Developing a Vision and Strategy:** The vision was not to be a top-down imposition. With inclusivity in mind, the coalition sought input from all corners of the company. Embracing the principle of giving change its due time, the company was patient, setting a realistic timeline of up to two years to fully see their vision come to fruition.
4. **Communicating the Change Vision:** True to the principle of transparency, communication was multifaceted. From town-hall meetings in Berlin's tech hubs to interactive digital platforms and dedicated internal podcasts, the change vision was disseminated in ways that resonated with every team member.
5. **Empowering Broad-Based Action:** Understanding the nuanced emotional responses to change, leadership emphasized creating safe spaces for dialogue. Workshops and feedback sessions became routine, allowing employees to voice concerns, reminisce about former methodologies, or express their hopes for the future.
6. **Generating Short-Term Wins:** In the sprawling urban landscape of Berlin, where tech successes are celebrated fervently, CyberGuardTech ensured that every milestone on their change journey was acknowledged. From improved team collaborations to enhanced client feedback, every win was a testament to the collective effort of the company.
7. **Consolidating Gains and Producing More Change:** While celebrating the vanguards of change, the early adopters, the company also laid out clear strategies to bridge the chasm to the early majority. Tailored training sessions, mentorship programs, and peer-led workshops became instrumental in ensuring that the change wasn't just for the few but was embraced by the many.

8. **Anchoring New Approaches in the Culture:** Berlin's dynamic culture, always in flux, yet rooted in its history, became an apt metaphor for CyberGuardTech's approach to anchoring change. The company initiated periodic reviews and feedback loops, ensuring the change was not a transient phase but became an integral part of the company's ethos.

As this transformative journey unfolded, the leadership at CyberGuardTech GmbH bore witness to the potency of their dual approach. Processes and strategy, while imperative, had to be harmoniously blended with understanding the human element – the fears, aspirations, and behaviors of their team. It was a holistic approach, encapsulating the structured and the organic, ensuring not just a change but a lasting evolution.

CHALLENGES AND TRIUMPHS

Navigating the path to meaningful transformation, CyberGuardTech GmbH faced its fair share of adversities. Yet, these challenges were matched, if not surpassed, by moments of celebration and realization of the promising direction in which the company was headed.

Challenges

1. **Resistance to Change:** Like any organization trying to pivot, CyberGuardTech GmbH faced internal resistance. Some employees, especially those who had been with the company since its inception, found it difficult to break away from old practices and routines.
2. **Communication Gaps:** Despite their best efforts to maintain transparency, there were moments when messages got lost in translation, leading to misconceptions and further fueling the fires of resistance.
3. **Resource Constraints:** The move toward an inclusive, widespread change demanded not only time and effort but also financial resources. Balancing the budget while ensuring that no aspect of the transformation journey was overlooked posed a significant challenge.
4. **External Market Pressures:** In the competitive tech ecosystem of Berlin, the market waits for no one. Even as CyberGuardTech GmbH was in its metamorphic phase, they had to fend off aggressive competitors and meet the evolving expectations of their clients.
5. **Balancing the Old and New:** Juggling existing contracts and obligations while trying to usher in a new era for the company was not without its complications. It was a tightrope walk, ensuring that neither the present nor the future was compromised.

Triumphs

1. **Inclusivity in Decision-Making:** One of the most celebrated achievements was the successful creation of the diverse guiding coalition, which became a beacon for collective decision-making and was a testament to the company's commitment to inclusivity.
2. **Rekindled Client Relationships:** As the change began to take shape, clients who had previously been wary started to notice and appreciate the enhanced service quality and innovative solutions that CyberGuardTech GmbH was bringing to the table.
3. **Employee Engagement:** The platform for emotions, paired with regular feedback sessions, led to a surge in employee engagement. The workforce began to feel more valued, leading to higher productivity and a stronger bond with the company's vision.

4. **Short-Term Milestones:** With every small victory, be it a successful project completion under the new methodology or positive internal survey results, CyberGuardTech ensured these wins were highlighted and celebrated, boosting morale and reinforcing the belief in the transformative journey.
5. **Market Recognition:** Beyond the internal accolades, the external world began to take notice. CyberGuardTech GmbH started receiving recognition in Berlin's tech circles, getting nominated for innovation awards, and being cited in tech publications as an example of a company that had successfully intertwined leadership principles with structured change management.

The journey of CyberGuardTech GmbH was punctuated by both stumbling blocks and moments of exultation. But with every challenge, the company's commitment to its vision and its unwavering dedication to blending structured change with emotional intelligence and leadership acumen made the transformation not only possible but exemplary.

ANCHORING THE CHANGE

In the dynamic landscape of IT security, changes can be fleeting. But for CyberGuardTech GmbH, the goal was never about mere surface-level transformations. They aimed to embed these shifts deep within their corporate DNA, ensuring that the organization didn't just change but evolved. Here's how they anchored their transformation for long-term success:

1. **Revisiting Company Values:** Central to CyberGuardTech GmbH's anchoring process was a re-evaluation of its foundational values. The leadership team, in collaboration with employees from all levels, sculpted a new set of core values that resonated with the transformed vision of the company. These values became the guiding principles for all future decisions and actions.
2. **Continuous Learning & Development:** Recognizing that change is continuous, and the tech landscape is always evolving, the company invested heavily in training programs. These programs were designed to instill new methodologies while fostering a culture of perpetual learning and adaptation.
3. **Feedback Mechanisms:** Periodic feedback sessions became a norm. These sessions were not limited to performance reviews but extended to understanding the employees' alignment with the company's new direction and addressing any discomforts or gaps.
4. **Celebrating Success Stories:** CyberGuardTech GmbH made it a point to consistently highlight and celebrate success stories that epitomized the new changes. Whether it was an individual's achievement or a team success, it was showcased, ensuring that the entire organization recognized and aspired to the standards set by these successes.
5. **Reinforced Leadership Commitment:** The company's top brass ensured that they were the flagbearers of the change. By embodying the new principles in their actions and decisions, they set a clear precedent for the rest of the organization to follow.
6. **Integration into Onboarding Processes:** To ensure that new employees were immediately aligned with the company's transformed ethos, the onboarding processes were revamped. Newcomers were introduced to the company's journey of change, its significance, and their role in this evolved narrative.

7. **Client & Stakeholder Engagement:** CyberGuardTech GmbH took proactive measures to communicate its transformation journey to its clients and stakeholders. By showcasing their commitment to this new path, they not only garnered external support but also solidified their internal resolve to stay on course.
8. **Regular Review and Iteration:** Recognizing that no change is perfect, the leadership instituted regular reviews of the transformation process. These reviews ensured that the company remained agile, making tweaks and iterations as and when required, ensuring that the transformation was always in line with the company's overarching goals.

By intertwining these steps into their operational fabric, CyberGuardTech GmbH didn't just implement change; they lived it. Over time, the changes moved from being "new initiatives" to just "the way things are done here." The company emerged not only as a leading IT security solution provider in Berlin but also as a testament to the power of anchored, purpose-driven transformation.

INTEGRATING KPIs FOR TRANSFORMATION MEASUREMENT

During business transformation, the importance of Key Performance Indicators (KPIs) cannot be overstated. These indicators, a blend of both hard and soft metrics, serve as navigational beacons, guiding and assessing the progress of organizational change. For a comprehensive evaluation, it's crucial to balance these KPIs, ensuring they align with the organization's goals, the transformation's focus, and its stage in the change process.

Balancing Hard and Soft KPIs

1. Hard KPIs
    - **Financial Performance (Revenue Growth, Cost Reduction):** These are quintessential for measuring the direct impact of changes on the bottom line.
    - **Operational Metrics (Market Share, Customer Acquisition and Retention Rates, Operational Efficiency):** Essential for evaluating market position, customer base growth, and efficiency improvements.
2. Soft KPIs
    - **Employee Metrics (Satisfaction, Engagement, Innovation Index):** These provide insights into how changes are affecting employee morale and fostering a culture of innovation.
    - **Adaptability Metrics (Change Adoption Rate, Leadership Effectiveness):** Crucial for understanding how smoothly the organization is transitioning and how effectively its leaders are facilitating this change.
    - **Customer Perspective (Client Satisfaction):** Reflects how external stakeholders perceive the changes and their impact on customer service and product quality.

Customizing the KPI-Mix

- The ratio of hard to soft KPIs should be tailored to the organization's specific transformation goals. For a technology company like CyberGuardTech GmbH, undergoing a significant operational and cultural shift, a balanced approach might be ideal.
- During the initial stages, soft KPIs can dominate to gauge immediate reactions to change initiatives. As the transformation matures, a gradual shift toward hard KPIs can help in measuring long-term impacts.

Implementing and Reviewing KPIs

- KPIs should be regularly reviewed for their relevance and effectiveness. This dynamic approach allows for adjustments based on real-time feedback and changing objectives.
- Engaging different departments in selecting and reviewing KPIs ensures a holistic perspective, accounting for the diverse impacts of transformation across the organization.

Integrating a balanced mix of hard and soft KPIs is crucial to get a comprehensive view of an organization's transformation progress. The ideal ratio of hard to soft KPIs depends on several factors, including the nature of the business, the specific goals of the transformation, and the stage of the transformation process. Here's a general approach to combining these KPIs:

1. **Assess Organizational Goals and Context**
    - If the transformation is primarily aimed at financial improvement or market expansion, a heavier emphasis on hard KPIs might be appropriate.
    - If the focus is on organizational culture change, employee engagement, or customer satisfaction, then soft KPIs should take precedence.
2. **Consider the Transformation Stage**
    - In the early stages, soft KPIs might be more relevant as they can provide immediate feedback on employee and customer reactions to the change.
    - As the transformation progresses, hard KPIs become crucial to assess the tangible impacts of the changes implemented.
3. **Balancing for a Holistic View**
    - A balanced approach could typically involve a 60-40 or 50-50 split between hard and soft KPIs. For instance, for every three hard KPIs, include two soft KPIs or maintain an equal number of both.
    - This balance ensures that while financial and operational targets are met, the human element of change is not neglected.
4. **Customizing Based on Feedback and Data**
    - Regularly review the effectiveness of the KPIs used. If certain metrics consistently fail to provide insightful data, they may need to be adjusted or replaced.
    - Be open to shifting the ratio based on periodic reviews. For example, if initial soft KPIs indicate low employee morale, you might focus more on soft KPIs in the short term to address this issue.
5. **Incorporating Industry-Specific Considerations**
    - Certain industries might inherently lean more toward either hard or soft KPIs. For instance, in service-oriented sectors like hospitality, soft KPIs related to customer satisfaction may be more heavily weighted.

A Case in Point – CyberGuardTech GmbH

- In the context of CyberGuardTech GmbH, the leadership team carefully selected KPIs that resonated with their specific challenges and objectives.
- For instance, to address the technological stagnation, they prioritized KPIs around revenue growth from new product lines and customer acquisition rates in emerging markets.
- Simultaneously, employee engagement surveys and innovation indices were employed to ensure the workforce remained motivated and creatively involved.

Hard KPIs

1. **Revenue Growth**
   - *Explanation:* This measures the increase in an organization's income over time, indicating market success and financial health.
   - *Example:* A software company aims for a 20% increase in annual revenue following the launch of a new product line.
2. **Market Share**
   - *Explanation:* This KPI tracks the percentage of an industry's sales that a particular company controls, reflecting competitive positioning.
   - *Example:* A retail chain monitors its market share growth post-rebranding to assess its standing against competitors.
3. **Customer Acquisition and Retention Rates**
   - *Explanation:* These metrics gauge how effectively a company attracts and retains customers, signifying the appeal and value of its offerings.
   - *Example:* A telecom company sets targets to increase new customer sign-ups by 15% and improve retention rates by 10%.
4. **Operational Efficiency**
   - *Explanation:* This measures the effectiveness of a company's operations, usually in terms of cost, time, and resources.
   - *Example:* A manufacturing firm aims to reduce production time by 25% while maintaining quality.
5. **Cost Reduction**
   - *Explanation:* This KPI tracks the decrease in operational or production costs, contributing to improved profitability.
   - *Example:* An e-commerce company targets a 20% reduction in logistics costs through process optimization.

Soft KPIs

1. **Employee Satisfaction and Engagement**
   - *Explanation:* These metrics reflect how happy and committed employees are in their roles, which impacts productivity and retention.
   - *Example:* A consulting firm conducts bi-annual surveys to measure employee satisfaction, aiming for a year-over-year improvement.
2. **Innovation Index**
   - *Explanation:* This KPI measures the rate and effectiveness of innovation within an organization, indicating adaptability and growth potential.
   - *Example:* A tech startup tracks the number of new patents filed annually as a measure of its innovation.
3. **Change Adoption Rate**
   - *Explanation:* This metric assesses how quickly and effectively a new system, process, or policy is embraced by the organization.
   - *Example:* A healthcare provider monitors the adoption rate of a new digital record-keeping system by its staff.
4. **Client Satisfaction**
   - *Explanation:* These metrics assess how satisfied customers are with a company's products or services, a critical factor in driving business growth.
   - *Example:* A hotel chain uses customer feedback scores to gauge satisfaction levels post a service revamp.

5. **Leadership Effectiveness**
   - *Explanation:* This KPI evaluates the impact of leadership on organizational performance and morale.
   - *Example:* A non-profit organization uses 360-degree feedback to assess the effectiveness of its management team.

CONCLUSION

The metamorphosis of CyberSecure GmbH is not just a tale of strategic shifts or operational overhauls; it's a narrative of resilience, foresight, and unwavering commitment to a vision. Looking back on this transformative journey reveals a myriad of lessons, each invaluable in its own right.

Firstly, **change is never a solitary endeavor.** While leadership can set the direction, it's the collective spirit of every member of the organization that propels it forward. CyberSecure GmbH's success was rooted in its ability to foster a culture of collaboration, where every voice mattered and every perspective contributed to the broader vision.

The importance of grounding change in core values also cannot be overstated. These foundational beliefs acted as the company's North Star, ensuring that in the face of challenges and uncertainties, there was a clear, unwavering point of reference to return to.

Another crucial insight was the **interplay between patience and persistence.** Transformational journeys are marathon endeavors, not sprints. While the urgency of change is undeniable, it's equally important to recognize that lasting change needs time to take root, to be internalized, and to manifest in tangible results. The leadership of CyberSecure GmbH demonstrated the ability to strike this delicate balance, driving the company forward while allowing space and time for organic growth and alignment.

The journey also highlighted the **significance of adaptability.** In a domain as dynamic as IT security, the only constant is change itself. CyberSecure's proactive approach to continuous learning and iterative improvements ensured they were always a step ahead, ready to embrace the next wave of change.

Lastly, this transformative odyssey underscored the fact that **change isn't just about the end goal.** The process, with its highs and lows, challenges and triumphs, is equally significant. It's in navigating these intricacies that organizations discover their true strengths, build character, and forge bonds that stand the test of time.

In essence, the story of CyberSecure GmbH serves as a testament to the power of visionary leadership, the strength of collective endeavor, and the transformative potential that lies in embracing change not as a disruption, but as an opportunity for evolution and growth. As they continue to thrive in Berlin's competitive IT security landscape, their journey offers inspiration and insights for organizations worldwide, reminding us that with the right mindset, strategies, and commitment, change can indeed be the harbinger of unmatched success.

CYBER CHUCKLES: REBOOTING LEADERSHIP – THE ULTIMATE PATCH FOR CHANGE MANAGEMENT

In the digital realm where change is the only constant, imagine leadership as the ultimate software patch, seamlessly integrating with the framework of change management. It's here, in the nexus of innovation and adaptation, where leaders morph into cyber-sherpas, guiding their teams through the ever-evolving landscape of IT security.

The fusion protocol: Leadership meets change management

Enter the fusion protocol, where the art of leadership blends with the science of change management. Picture a scenario where the command line of leadership inputs intersects with the algorithms of Kotter's Eight-Step Change Model, creating a dynamic, adaptive system ready to take on the cyber world's unpredictability.

1. **Creating Urgency with Vision:** Imagine igniting the digital spark in every team member, crafting a vision so compelling that the status quo feels like a dial-up connection in a fiber-optic world.
2. **Forming a Cyber Coalition:** Here, the coalition isn't just the C-suite. It's a band of digital disruptors and innovators from every corner of the organization, each armed with unique codes and insights, ready to champion the change.
3. **Strategic Command Lines:** In this chapter, setting a vision for change is akin to writing a master script, one that runs seamlessly across departments, rallying the troops toward a common goal with clarity and purpose.
4. **Broadcasting the Vision:** Think of it as the ultimate social engineering feat, but for good. Communicating the vision becomes a campaign, spreading like a beneficial virus, infecting the organization with motivation and drive.
5. **Debugging the Path:** Leadership in this domain means identifying and removing barriers to change, much like a seasoned programmer debugs code, ensuring a clear path forward.
6. **Celebrating Milestones:** Every small win is a level up, a moment to pause, celebrate, and recharge before diving back into the digital fray, boosting morale, and reinforcing the journey's value.
7. **Iterative Development:** Building on the momentum, this phase is about iterating, enhancing, and expanding on the successes, much like software updates that add features and improve security.
8. **Embedding the Change:** The final step in our fusion protocol is ensuring that change isn't just a temporary patch but becomes part of the organization's core operating system, ingrained in its culture and practices.

Epilogue: Beyond the binary

As the final lines of code in this chapter compile, it's clear that leading through change in the cyber world transcends traditional binaries. It's about being fluid yet focused, innovative yet inclusive, commanding yet collaborative. In this landscape, a leader doesn't just direct; they inspire, empower, and co-navigate the journey of transformation, turning challenges into steppingstones for a resilient, future-proof organization.

So, as we log off from this chapter, remember that in the cyber age, leadership combined with change management isn't just about steering the ship. It's about upgrading it, ensuring it's swift, agile, and ready to sail the digital seas, no matter how rough the waters.

Chapter 4

The security mindset paradox

INTRODUCTION

- **The Curious Case of Jane:** Why an IT pro might use "Password123" on a critical system.
- **The Password Manager's Lament:** A tech perspective on user habits.
- **Convenience vs. Security:** The tug-of-war we all face daily.

Unreliable Narrator: Meet Jane, an IT professional logging into a critical system with "Password123." You wonder why she, of all people, would make such a rookie mistake. As it turns out, Jane is a security tester, deliberately exploiting human vulnerabilities to demonstrate the importance of robust security measures.

IMPACT AND COST

Financial Costs: Data breaches aren't just technical headaches – they're financial nightmares. In 2020, the average cost of a data breach was $3.86 million. Beyond direct costs, breaches erode trust and damage reputations, sometimes irreparably. Equifax's 2017 breach, for example, resulted in a $700 million settlement and lasting reputational damage.

Real-World Examples: The WannaCry ransomware attack in 2017 exploited known vulnerabilities. Despite available patches, many organizations failed to update their systems, leading to widespread disruption and financial loss.

Fear Appeals: The emotional consequences of breaches are severe. Identity theft can turn lives upside down, leading to financial loss and the painstaking process of reclaiming one's identity. Imagine the inconvenience of disputing fraudulent charges, the frustration of dealing with compromised accounts, and the fear of personal information being misused.

COMMON COGNITIVE BIASES

Overconfidence Bias: Believing our systems are more secure than they are. Example: An IT team assumes their network is impregnable, neglecting regular updates.

Anchoring Bias: Relying too heavily on past experiences. Example: A company that's never been breached assumes they're not a target, ignoring emerging threats.

Confirmation Bias: IT teams might ignore early warning signs of a breach because they believe their security measures are infallible.

Cognitive Dissonance: Employees experience discomfort when their actions (using weak passwords) contradict their knowledge (importance of strong passwords).

DOI: 10.1201/9781003562184-7

Habituation: Over time, employees may become desensitized to security warnings.

Real-World Examples: The Target breach in 2013 was partly due to overconfidence in their security measures, leading to insufficient monitoring of their network.

Shift the focus: The password manager's lament

From the perspective of a password manager app: "Hello, I'm your trusty password manager. Every day, I watch you create passwords like '123456' and 'Password123.' Despite my best efforts to suggest stronger options, convenience often wins. It's a struggle to get users to see the importance of robust security. But oh, the stories I could tell of those who didn't listen ..."

Transition: But what happens when these warnings are ignored? Let's look at some real-world consequences ...

STRATEGIES TO OVERCOME BIASES

Addressing Overconfidence Bias

Regular Security Assessments: Conduct frequent, comprehensive security assessments.

Simulations: Use breach simulations to expose vulnerabilities and stress-test systems.

Combating Anchoring Bias

- **Continuous Learning:** Stay updated with the latest threats and security technologies.
- **Dynamic Policies:** Adapt security policies based on current threat landscapes, not just past experiences.

Addressing Confirmation Bias

- **Third-Party Audits:** Regular third-party audits can provide an objective assessment of security measures.
- **Red Teaming Exercises:** Simulated attacks by red teams can uncover vulnerabilities that internal teams might overlook.

Reducing Cognitive Dissonance

- **Simplify Secure Practices:** Make it easy for employees to follow secure practices, reducing the gap between knowledge and action.

Combating Habituation

- **Refresh Training Materials:** Regularly update and rotate training methods to keep security messages fresh.
- **Gamification:** Use gamification to make security practices more engaging and less routine.

Leveraging the Default Effect

- **Secure Defaults:** Configure systems to enforce complex passwords, automatic logouts, and other secure settings by default.

Loss Aversion: People are more likely to avoid losses than pursue gains. Frame security practices as protecting something valuable (data, privacy) rather than just following rules.

Mental Fatigue: Our ability to make good decisions weakens throughout the day. Highlight the importance of strong default security settings (e.g., password managers) to combat decision fatigue.

Social Proof: People are more likely to follow the crowd. Emphasize the importance of security awareness training that normalizes secure behavior within the organization.

Fear Appeals: The inconvenience of dealing with data breaches, such as recovering from identity theft or handling financial losses, can create a sense of urgency.

PLAY WITH TIME: FROM BREACH TO DECISION

Open with a Data Breach: Imagine waking up to news of a major data breach. Rewind six months: security assessments were skipped, updates postponed, and convenience prioritized. Fast forward: the breach's aftermath shows the long-term consequences of these choices.

Transition: This timeline highlights the ripple effect of security decisions. But how do we address these biases and improve security culture?

Role of leadership

Leadership Styles

- **Authoritarian Leadership:** Strict enforcement of security policies can lead to compliance but may stifle creativity.
- **Transformational Leadership:** Inspires employees to prioritize security through motivation and engagement.

Actionable Steps

- **Incentivizing Secure Practices:** Implement reward programs for employees who demonstrate good security behaviors.
- **Training Programs:** Regular, engaging training sessions to keep security at the forefront of employees' minds.

Sensory overload: The password dilemma

Imagine the internal struggle when creating a new password:

- **Voice of Convenience:** "Just use 'Password123' again. It's easy!"
- **Voice of Security:** "No, it has to be strong and unique!"
- **Pop-up Warnings and Meters:** "Weak password! Try again!" "Strong password recommended."

Transition: This internal battle highlights the constant tug-of-war between convenience and security. But what are the long-term consequences of prioritizing convenience? Let's explore the impact of data breaches...

PSYCHOLOGICAL TACTICS IN CYBERATTACKS

Phishing Attacks: Beyond technical details, phishing emails often create a sense of urgency or exploit trust. Example: An email claiming your bank account has been compromised urges immediate action, leading to hurried, less secure decisions.

Social Engineering: Attackers exploit human psychology to gain access to information or systems. Example: Pretending to be an IT support staff member, an attacker convinces an employee to divulge their login credentials.

Bystander Effect: People are less likely to intervene if they believe others will. Example: Multiple employees noticing suspicious activity but assuming someone else will report it. Discuss the importance of creating a culture of security where everyone feels responsible for reporting suspicious activity.

EXAMPLES OF SUCCESS

Case Studies

- **Company A:** Overcame the paradox by implementing a robust security training program and rewarding employees for identifying security vulnerabilities.
- **Company B:** Used transformational leadership to foster a culture where security is everyone's responsibility. Leaders led by example, and employees were engaged through interactive security drills.

Analysis: Effective strategies included leadership commitment, continuous training, and positive reinforcement.

Transition: These success stories show that change is possible. Now, let's bring it all together ...

CONCLUSION

Recap Key Strategies: Overcoming the security mindset paradox requires addressing cognitive biases through tailored strategies, strong leadership, and continuous engagement.

Recommendations

- **Foster a Culture of Security:** Encourage open discussions about security, reward secure behaviors, and ensure leadership sets the tone.
- **Continuous Improvement:** Regularly update security practices and policies to adapt to new threats.
- **Employee Engagement:** Make security training engaging and relevant to keep employees invested in maintaining robust security practices.

By bridging the gap between knowing and doing, organizations can cultivate a culture of proactive security awareness, reducing vulnerabilities and strengthening their defenses against evolving cyber threats.

CYBER CHUCKLES

- **Why did the hacker only use weak passwords?** Because strong ones took too long to type after a stressful day!
- **How many IT pros does it take to change a password?** None, if it's "Password123" – why change perfection?

- **Convenience vs. Security:** "I know I should lock my door, but who has the time for keys?"
- **Why did the security team cross the road?** To change the default password on the other side!
- **How do you catch a cybercriminal?** With strong passwords and a net full of common sense!
- **What's a hacker's favorite season?** Phishing season!

Chapter 5

Psychological principles to boost cybersecurity awareness

Spoiler Alert!

- This chapter delves into the world of behavioral science to make people care about cybersecurity (because apparently, we all need a little nudge sometimes).
- Learn about cool tricks like "nudging" and "gamification" to make security awareness campaigns more engaging (who knew learning about firewalls could be fun?).

INTRODUCTION

This chapter presents an innovative strategy to enhance cybersecurity awareness by embedding psychological principles into training programs. In the face of continually evolving cyber threats and the existing gap in effective cybersecurity behaviors, there's a pressing need for fresh approaches. This approach aims to deepen understanding, engagement, and behavioral change among users, thereby strengthening the human element in cyber defense. Digital interconnectedness spans both personal and professional spheres, cybersecurity extends beyond traditional confines, becoming essential for both individuals and organizations. As cyber threats grow more complex and sophisticated, securing digital assets becomes increasingly challenging. However, the key to effective cybersecurity resides in addressing the human factor alongside technological solutions. This white paper delves into how integrating psychological insights into cybersecurity strategies can significantly boost awareness and cultivate a culture of responsible digital conduct, empowering users to make informed choices and build a solid defense against evolving cyber threats.

SELECTING THE FIVE PRINCIPLES

Five psychological principles have been chosen for this whitepaper based on their proven effectiveness in influencing human behavior and their relevance to cybersecurity awareness:

1. **Behavioral Economics (Nudging):** Nudging involves subtle alterations in the environment to guide individuals toward desired behaviors without restricting their choices.
2. **Social Learning Theory**: Social learning theory emphasizes the role of observation and imitation in acquiring new behaviors.
3. **Gamification:** Gamification applies game-like elements to non-game contexts to enhance motivation, engagement, and learning.
4. **Theory of Planned Behavior (TPB):** The TPB suggests that intentions are the strongest predictor of behavior, and it identifies three key factors that influence intentions: attitudes, subjective norms, and perceived behavioral control.

 DOI: 10.1201/9781003562184-8

5. **Cognitive Behavioral Therapy (CBT):** CBT focuses on identifying and modifying negative thought patterns and behaviors that contribute to maladaptive behaviors, such as clicking on phishing links or failing to maintain strong passwords.

These principles offer a diverse range of approaches to addressing the psychological factors that influence cybersecurity behaviors. By understanding and applying these principles, organizations and individuals can develop more effective cybersecurity awareness programs and strategies.

SECTION I: BEHAVIORAL ECONOMICS (NUDGING)

Introduction to nudging

Nudging, a concept rooted in behavioral economics, leverages subtle cues and environmental adjustments to influence decision-making and behavior. Operating under the premise that human choices are often shaped by the design of the choices presented, nudging aims to steer individuals toward beneficial behaviors without restricting their freedom of choice. In the context of cybersecurity, nudging can be a potent tool, subtly guiding users toward more secure behaviors by capitalizing on innate cognitive biases and decision-making shortcuts.

Real-world example

Consider the implementation of nudging in software installation processes. Often, the most secure settings are not the default, leading users to inadvertently opt for less secure options. A nudge in this scenario would be setting the most secure option as the default choice, subtly encouraging users toward better security practices without limiting their ability to choose otherwise.

Benefits and disadvantages

Benefits: Nudging is a subtle, non-intrusive method that can effectively shift user behavior toward enhanced security practices. It respects user autonomy while gently guiding choices.

Disadvantages: Ethical considerations arise around nudging, as it could be perceived as manipulative. It's crucial to maintain transparency and ethical standards, ensuring that nudges are designed with the user's best interests in mind and do not compromise informed decision-making.

Cybersecurity awareness application

Nudging can be applied to various aspects of cybersecurity awareness, such as:

- Default settings that favor security, like automatic updates or strong password requirements.
- Designing user interfaces that highlight secure options or provide timely security reminders.
- Incorporating security prompts and warnings that are easy to understand and act upon.

Implementation ideas

Implementing nudging effectively requires:

- A thorough understanding of user behaviors and common cybersecurity pitfalls.
- Designing interfaces and processes that naturally lead to more secure behaviors.
- Regular assessment and refinement of nudges based on user feedback and evolving cybersecurity landscapes.

SECTION 2: SOCIAL LEARNING THEORY

Introduction to the theory

Social Learning Theory, pioneered by Albert Bandura, posits that behavior is learned through observation, imitation, and modeling. This theory underscores the significance of social influence and observational learning in shaping individual behavior. In the realm of cybersecurity, social learning theory suggests that individuals are more likely to adopt secure online practices when they observe these behaviors being valued and practiced by their peers, leaders, or influential figures within their community or organization.

Real-world example

A practical application of social learning theory can be seen in workplace training programs. For instance, when a company's leadership actively participates in cybersecurity training and openly discusses their security practices, employees are more likely to view these behaviors as normative and important, leading to increased adoption of similar practices. This effect is amplified when leaders share their experiences of thwarting cyber threats or implementing effective security measures.

Benefits and disadvantages

Benefits: Social learning theory can create a positive ripple effect, fostering a culture of cybersecurity awareness and practice. It harnesses the power of social influence and peer behavior to reinforce learning and adoption of secure practices.

Disadvantages: The theory relies heavily on the presence of role models or influential figures exhibiting the desired behavior. In the absence of such figures, or if they display poor cybersecurity habits, the theory's effectiveness can be diminished.

Cybersecurity awareness application

Applications of Social Learning Theory in cybersecurity awareness can be diverse:

- Encouraging leaders and influencers within the organization to model robust cybersecurity practices.
- Creating peer-led cybersecurity training sessions where employees share best practices and experiences.
- Developing mentorship programs where experienced staff guide others in adopting secure online behaviors.

Implementation ideas

Effective implementation of Social Learning Theory in cybersecurity initiatives requires:

- Identifying and empowering internal cybersecurity champions to act as role models.
- Encouraging open dialogue and sharing of experiences related to cybersecurity within the organization.
- Recognizing and rewarding good cybersecurity practices to reinforce positive behavior.
- Incorporating storytelling and real-life examples in cybersecurity training to enhance relatability and engagement.

SECTION 3: GAMIFICATION

Introduction to gamification

Gamification involves the integration of game-like elements and dynamics into non-gaming contexts, such as education and training. This approach capitalizes on the innate human desire for play, competition, achievement, and status to increase engagement and motivation. In cybersecurity awareness, gamification can transform traditional, often mundane, security training into a more engaging, interactive, and memorable experience. By doing so, it not only enhances learning but also encourages the application of cybersecurity best practices in a fun and engaging manner.

Real-world example

A notable example of gamification in cybersecurity is the creation of interactive learning platforms where users engage in simulated cyber threat scenarios. These platforms often include elements like point scoring, leaderboards, achievement badges, and progress tracking. For instance, an organization might develop a security challenge where employees earn points or badges for completing security quizzes, identifying phishing emails, or reporting potential security threats, thereby promoting active participation in cybersecurity practices.

Benefits and disadvantages

Benefits: Gamification can significantly increase engagement and motivation in cybersecurity training, leading to higher retention of information and better application of learned principles. It can also foster a positive attitude toward cybersecurity, making it less intimidating and more accessible.

Disadvantages: The challenge lies in designing gamification elements that are relevant and meaningful, rather than just superficially entertaining. Overemphasis on competition can sometimes overshadow the learning objectives, and not all individuals may respond equally to gamified elements.

Cybersecurity awareness application

Gamification can be applied to various aspects of cybersecurity awareness, such as:

- Developing interactive and competitive cybersecurity training modules.
- Incorporating quizzes, puzzles, and challenges related to cybersecurity into the workplace environment.
- Establishing reward systems for recognizing and celebrating cybersecurity achievements and behaviors among staff.

Implementation ideas

For effective implementation of gamification in cybersecurity awareness:

- Identify key learning objectives and ensure that gamification elements align with these goals.
- Design inclusive and varied gamification features to cater to different learning styles and preferences.

- Regularly update and refresh gamification content to maintain engagement and relevance.
- Balance competition with collaboration, ensuring that the gamified elements encourage a supportive and inclusive cybersecurity culture.

SECTION 4: THEORY OF PLANNED BEHAVIOR

Introduction to the theory

The TPB, developed by Icek Ajzen, posits that an individual's intention to engage in a behavior is the most significant predictor of that behavior. This intention is influenced by three key factors:

1. **Attitudes (Personal Beliefs about the Behavior)**
   Attitudes refer to an individual's positive or negative evaluations of a behavior. A positive attitude toward cybersecurity behaviors, such as practicing strong password hygiene and avoiding phishing scams, increases the likelihood of an individual engaging in those behaviors.
2. **Subjective Norms (Perceived Social Pressures)**
   Subjective norms refer to an individual's perception of the social pressure to perform or not perform a behavior. If an individual perceives that others view cybersecurity behaviors as important, they are more likely to adopt those behaviors themselves.
3. **Perceived Behavioral Control (the Perceived Ease or Difficulty of Performing the Behavior)**
   Perceived behavioral control refers to an individual's belief in their ability to perform a behavior. If an individual believes that they have the skills and resources to protect their cybersecurity, they are more likely to take proactive measures.

In cybersecurity, TPB can be utilized to understand and shape employees' intentions toward engaging in secure online behaviors, thereby enhancing overall cybersecurity practices.

Real-world example

An application of TPB in a cybersecurity context could be seen in an organization's approach to password management. By assessing employees' attitudes toward the importance of strong passwords, understanding the social norms within the workplace around password security, and addressing perceived barriers (like the inconvenience of frequently changing passwords or using complex password formulas), training programs can be tailored to positively influence employees' cybersecurity behaviors.

Benefits and disadvantages

Benefits: TPB provides a comprehensive framework for understanding the factors that influence cybersecurity behavior, making it easier to design targeted interventions. It considers the multifaceted nature of behavior change, addressing not just knowledge but also social influences and perceived control.

Disadvantages: The theory requires a deep understanding of the audience's attitudes, beliefs, and perceptions, which can be challenging to accurately gauge. Additionally, it may not fully account for habitual or impulsive behaviors that occur outside conscious decision-making processes.

Cybersecurity awareness application

The TPB can be applied to cybersecurity awareness initiatives in various ways, such as:

- Surveying employees to understand their attitudes, subjective norms, and perceived control regarding cybersecurity.
- Tailoring communication and training materials to address identified attitudes and beliefs.
- Leveraging social influence by highlighting positive cybersecurity behaviors among peers and leadership.
- Enhancing perceived behavioral control by providing necessary tools and resources to perform secure online actions easily.

Implementation ideas

To implement TPB effectively in a cybersecurity context:

- Conduct comprehensive assessments to understand employees' current perceptions and behaviors regarding cybersecurity.
- Develop targeted interventions based on these insights, addressing specific attitudes, subjective norms, and perceived barriers.
- Regularly evaluate and adjust strategies based on feedback and changing cybersecurity landscapes.
- Conduct surveys or focus groups to gather data on attitudes, subjective norms, and perceived behavioral control among the target audience.
- Develop cybersecurity training materials that address common misconceptions and address the identified factors.
- Encourage collaboration and peer support to reinforce positive cybersecurity behaviors among employees.
- Emphasize the importance of cybersecurity through regular communication and reminders.

SECTION 5: COGNITIVE BEHAVIORAL THERAPY (CBT)

Introduction to CBT

CBT is a psychological treatment approach that focuses on identifying and altering dysfunctional thinking patterns, beliefs, and attitudes, which in turn can influence behavior. CBT is grounded in the understanding that negative thought patterns can lead to maladaptive behaviors. In the context of cybersecurity, CBT can be instrumental in identifying and modifying unhelpful beliefs and behaviors that lead to cybersecurity risks, such as careless handling of sensitive data or susceptibility to phishing attacks.

Real-world example

An example of CBT in cybersecurity is addressing the belief that "my actions won't make a difference in overall cybersecurity." Through CBT techniques, individuals can be guided to recognize and challenge this belief, replacing it with more constructive thoughts like "every secure action I take contributes to the overall safety of my organization." This shift in thinking can lead to more responsible cybersecurity behaviors. A concrete example of how CBT can be applied to cybersecurity awareness is in addressing the common misconception that clicking on a suspicious link is harmless. CBT could help individuals recognize and challenge this belief by providing them with information about the risks associated with clicking on such links and teaching them coping mechanisms to manage anxiety associated with avoiding them.

Benefits and disadvantages

Benefits: CBT can effectively change the way individuals perceive cybersecurity risks and their role in mitigating them. It can lead to long-lasting behavioral changes as it addresses the root cause of behaviors.

- It can be tailored to individuals' specific needs and challenges.
- It has been shown to be effective in reducing risky online behavior and improving overall cybersecurity awareness.

Disadvantages: CBT requires active participation and commitment from individuals, which can be a challenge in a workplace setting. It may also require more time and resources compared to other training methods.

- It requires motivation and commitment from the individual for effective change.
- It may not be as effective in addressing the root causes of cybersecurity risks, such as lack of knowledge or technical skills.

CYBERSECURITY AWARENESS APPLICATION

CBT can be applied to cybersecurity awareness initiatives in various ways, such as:

- Developing training modules that focus on recognizing and challenging cybersecurity-related misconceptions.
- Implementing workshops or sessions where employees can practice CBT techniques, such as cognitive restructuring, in the context of cybersecurity.
- Providing one-on-one counseling or group discussions for employees who have experienced cybersecurity incidents, helping them cope with and learn from these events.

Implementation ideas

For effective implementation of CBT in cybersecurity awareness:

- Collaborate with psychologists or trained professionals to develop CBT-oriented cybersecurity training and resources.
- Ensure that CBT techniques are tailored to address specific cybersecurity concerns and are applicable in a workplace setting.
- Monitor and evaluate the impact of CBT interventions on employees' cybersecurity behaviors and make necessary adjustments.

CONCLUSION

The exploration of psychological principles in the context of cybersecurity awareness is pointing toward a fundamental idea: the human element is as crucial to cybersecurity as technological defenses. This paper has delved into five key psychological principles – Behavioral Economics (Nudging), Social Learning Theory, Gamification, TPB, and CBT – each offering unique insights and strategies for enhancing cybersecurity awareness and behavior.

The integration of these principles presents a multidimensional approach to cybersecurity, moving beyond traditional training methods. By understanding and leveraging the nuances of human behavior and psychology, organizations can develop more effective, engaging, and sustainable cybersecurity awareness programs.

- **Behavioral Economics** emphasizes subtle environmental modifications to encourage secure behavior, respecting individual autonomy while guiding choices.
- **Social Learning Theory** underscores the impact of role models and social influence, highlighting the power of observational learning in adopting secure practices.
- **Gamification** introduces an element of play, competition, and achievement, transforming cybersecurity training into an engaging and interactive experience.
- **Theory of Planned Behavior** provides a framework for understanding the factors influencing cybersecurity intentions and actions, allowing for more targeted and effective interventions.
- **Cognitive Behavioral Therapy** offers a method to identify and modify detrimental cybersecurity beliefs and behaviors, fostering a more security-conscious mindset.

The synergy of these psychological principles with technical solutions could form a robust defense against cyber threats. Organizations and individuals are encouraged to adopt this multifaceted approach, combining technical prowess with an in-depth understanding of human psychology, to cultivate a resilient and proactive cybersecurity culture. This approach not only addresses the immediate challenges of cybersecurity but also lays the foundation for enduring behavioral change, ensuring a safer digital environment for all.

Leveraging psychological principles offers a promising path to strengthening cybersecurity awareness and behaviors. To embark on this journey, organizations could:

1. Conduct a needs assessment to identify specific cybersecurity behavior gaps.
2. Collaborate with psychology experts to tailor awareness programs.
3. Pilot these programs in small groups for feedback.
4. Roll out comprehensive, interactive training sessions organization wide.

CYBER CHUCKLE: THE MIND GAMES OF CYBERSECURITY

Imagine an engaging comic strip that humorously juxtaposes traditional cybersecurity training with innovative, psychology-driven methods. The first panel shows a bored employee half-asleep during a standard lecture on password management, with a monotone voice droning on in the background. The next few panels contrast this with lively and interactive scenes inspired by the psychological principles you've mentioned.

1. **Nudging for Better Choices:** A character faced with a decision between two doors: one labeled "Easy Path, Weak Passwords" and the other "Smart Choice, Strong Passwords." The smart choice door is slightly ajar, glowing invitingly, nudging the character toward it.
2. **Social Learning Theory in Action:** A popular, superhero-like figure at the office, admired by all, demonstrates how to spot phishing emails. Colleagues watch in awe and mimic the actions, turning into mini cybersecurity heroes themselves.

3. **Gamification Galore:** Employees partake in a virtual game show titled "The Great Cyber Chase," where they earn points for identifying security threats, competing on a leaderboard that's displayed prominently in the office.
4. **Theory of Planned Behavior Illustrated:** A scene where characters are weighing their options on a giant scale. On one side are positive attitudes, subjective norms, and perceived behavioral control, tipping the scale toward taking proactive cybersecurity measures.
5. **CBT for Cyber Strength:** A character stands in front of a mirror, practicing positive affirmations like "I will not fall for phishing scams" and "I am capable of protecting my digital presence," transforming from a cybersecurity novice into a confident expert.

The final panel zooms out to show a vibrant and engaged office, with employees actively participating in cybersecurity practices, laughing, and learning together, showcasing a transformed cybersecurity culture that's both effective and enjoyable.

Chapter 6

Building a cyber-resilient culture

INTRODUCTION

- **The Ransomware Reality:** How one innocent click can halt your operations.
- **Shared Responsibility:** Why cybersecurity isn't just the IT department's job.
- **Beyond Tech Defenses:** The human element in creating a resilient security culture.

Imagine this: A seemingly harmless email from a trusted vendor lands in an employee's inbox. Clicking the link triggers a ransomware attack, encrypting critical data and grinding operations to a halt. Panic sets in. The company has no incident response plan, employees are unsure who to contact, and valuable time is wasted while scrambling for a solution. This scenario, unfortunately, plays out too often, highlighting the critical need for a cyber-resilient culture.

KEY CHARACTERISTICS

A cyber-resilient culture goes beyond technical defenses. It fosters a shared responsibility for cybersecurity within an organization. Here are the key pillars:

- **Security Awareness:** Employees at all levels understand cybersecurity threats and their role in mitigating them. This means regular training and updates on the latest threats, ensuring everyone is equipped to recognize and respond to potential risks.
- **Preparedness:** The organization has a well-defined incident response plan that outlines roles, responsibilities, and communication protocols in case of an attack. This ensures that when an incident occurs, there is no confusion, and everyone knows their role in mitigating the impact.
- **Collaboration:** Open communication channels exist where employees feel empowered to report suspicious activity and seek help without fear of reprimand. This creates an environment of trust and encourages proactive reporting of potential issues.

PRACTICAL STRATEGIES: GO BEYOND THE LIST

Building a cyber-resilient culture requires sustained effort. Here's how to turn those key characteristics into action:

Security Awareness Training

- **Interactive Modules:** Don't settle for boring lectures. Use interactive modules, simulations, and real-world examples to keep employees engaged. Make the content relatable and relevant to their daily tasks.

DOI: 10.1201/9781003562184-9

- **Gamification:** Techniques like leaderboards and badges for completing training or identifying phishing attempts can further incentivize participation. This adds an element of fun and competition, making the training more engaging.
- **Flexible Delivery:** Online training platforms allow for flexibility and cost-effective delivery. Employees can complete training at their own pace and convenience, ensuring higher participation rates.

Incident Response Exercises

- **Tabletop Exercises:** Regularly conduct tabletop exercises to test your incident response plan. These exercises simulate real-life scenarios, allowing teams to practice their responses in a controlled environment.
- **Simulated Scenarios:** Simulate different attack scenarios, identify gaps in communication, and refine your response procedures. This helps in identifying weaknesses in the plan and improving overall preparedness.
- **Role Clarity:** Ensure everyone knows their roles and can react quickly and efficiently in a real crisis. Clear role definitions prevent confusion and ensure a coordinated response.

Open Communication Channels

- **Safe Reporting Spaces:** Create a safe space for employees to report suspicious activity through an anonymous reporting hotline or dedicated email addresses. This encourages employees to report without fear of repercussions.
- **Internal Knowledge-Sharing Forums:** Encourage open communication by actively listening to concerns and addressing them promptly. Foster a culture where knowledge sharing is encouraged and rewarded.

TECHNOLOGY AS A FORCE MULTIPLIER

Technology can be a powerful ally in building a cyber-resilient culture:

- **Online Training Platforms:** Utilize these platforms to deliver engaging security awareness modules and track employee progress. These platforms can also provide real-time updates and reminders.
- **Knowledge-Sharing Forums:** Allow employees to discuss security concerns and learn from each other's experiences. Forums can be moderated to ensure accurate information is shared.
- **Collaboration Tools:** Tools like instant messaging and video conferencing can facilitate communication during incident response exercises and real-world crises. These tools ensure that teams can communicate quickly and efficiently, regardless of location.

ROLE OF GAMIFICATION AND POSITIVE REINFORCEMENT

Gamification techniques can inject fun and competition into security awareness training:

- **Leaderboards:** Create a healthy sense of competition by displaying top performers. This motivates employees to participate and excel in training activities.
- **Points and Badges:** Award points for completing training modules or badges for identifying phishing attempts to reinforce positive security behaviors. This recognition can boost morale and encourage continued participation.

- **Thoughtful Implementation:** Gamification should be used thoughtfully to avoid trivializing the importance of cybersecurity. Ensure that the focus remains on learning and improving security practices.

BEHAVIORAL SCIENCE INTEGRATION

Nudge Theory

- **Subtle Influences:** Apply nudges to security awareness, like automatically enrolling employees in security training or pre-populating strong password requirements during account creation. These small changes can significantly improve security compliance.

Loss Aversion

- **Framing Security:** Frame security practices as protecting something valuable (data, privacy) rather than just following rules. People are more likely to adopt security measures when they understand the potential losses they are avoiding.

Social Proof

- **Social Learning:** Highlight successful security awareness campaigns that used social learning elements, like employee testimonials or gamified leaderboards. When employees see their peers engaging in secure behaviors, they are more likely to follow suit.

FOCUS ON FUTURE TECHNOLOGIES

The Rise of AI

- **Personalized Training:** Explore how AI-powered security awareness training can personalize learning experiences and identify knowledge gaps for individual employees. AI can adapt training content to meet the specific needs and skill levels of each employee.

Immersive Technologies

- **VR Simulations:** Discuss the potential of virtual reality (VR) simulations to create more realistic and engaging incident response training scenarios. VR can provide immersive, hands-on experience in a safe and controlled environment.

Cybersecurity Mesh Architecture

- **Collaborative Security:** Explain how a mesh architecture can foster a more collaborative and distributed approach to security, empowering employees to be active participants in threat detection and prevention. This approach decentralizes security efforts, making it more difficult for attackers to compromise the entire system.

HUMAN-CENTRIC SECURITY

Security Champions Program

- **Internal Ambassadors:** Highlight the value of a security champion program where enthusiastic employees act as internal ambassadors, promoting best practices and fostering a culture of security ownership. These champions can serve as role models and provide peer support.

Mental Health and Security

- **Connection to Vigilance:** Discuss the connection between mental fatigue and security decision-making. Emphasize the importance of breaks and a positive work environment to enhance employee vigilance. Stressed and overworked employees are more likely to make mistakes.

Security for All

- **Inclusive Training:** Advocate for security awareness programs that cater to diverse learning styles and abilities, ensuring everyone feels included and empowered to contribute to a secure environment. Tailor training materials to be accessible to all employees, regardless of their technical expertise.

REAL-WORLD EXAMPLES (BEYOND THE USUAL)

Case Studies

- **Innovative Approaches:** Include examples of organizations from different industries that have successfully implemented innovative approaches to building a cyber-resilient culture. Highlight the unique strategies that led to their success.

Security Awareness Campaigns

- **Creative Campaigns:** Showcase creative and engaging security awareness campaigns that went beyond traditional methods and achieved significant improvements in employee behavior. Use these examples to inspire new approaches within your own organization.

Lessons from Breaches

- **Analysis and Mitigation:** Analyze recent high-profile data breaches and discuss how a stronger security culture could have mitigated the impact. Learn from these incidents to strengthen your own security practices.

NOVEL PERSPECTIVES

Deception Fatigue and Security Awareness

- **The Problem:** Employees are bombarded with security warnings and phishing attempts, leading to deception fatigue where they become desensitized and less likely to identify genuine threats.
- **The Solution:** Focus on quality over quantity. Deliver targeted security awareness training based on real threats and employee roles. Emphasize the importance of healthy skepticism and encourage employees to take breaks to avoid burnout.

Security Gamification and Ethics

- **The Concern:** Overly gamified security awareness programs can trivialize cybersecurity and create a culture of competition over collaboration.
- **The Recommendation:** Use gamification elements strategically. Reward positive security behaviors like completing training or reporting suspicious activity. Focus on team-based challenges that promote collaboration and a sense of shared responsibility.

Cybersecurity and Physical Security Convergence

- **The Traditional View:** Cybersecurity and physical security are often seen as separate disciplines.
- **The New Perspective:** Cyberattacks often involve physical elements (e.g., social engineering, tailgating). Integrate physical security measures (e.g., access control, security awareness training for physical security staff) into your cyber-resilient culture strategy.

Security Champions: Beyond the Tech-Savvy

- **The Traditional Model:** Security champions are often technical experts who promote best practices.
- **The Inclusive Approach:** Expand the security champion program to include employees from diverse departments (e.g., HR, Legal, Marketing). This leverages their unique skills and perspectives to create a more comprehensive security culture.

The Power of Storytelling in Security Awareness

- **The Limitation of Lectures:** Traditional security awareness lectures can be dry and forgettable.
- **The Power of Narrative:** Utilize storytelling techniques to create engaging and relatable security awareness campaigns. Use real-world security incidents (anonymized) to showcase the impact of cyberattacks and the importance of vigilant behavior.

Adaptive Learning Platforms

- **The Need for Personalization:** Standardized training can be ineffective. Adaptive learning platforms use AI to personalize training content based on the learner's progress, strengths, and weaknesses.
- **Implementation:** Integrating adaptive learning platforms can ensure that each employee receives training that is most relevant to their role and current knowledge level.

Crisis Communication Plans

- **Importance of Clear Communication:** During a cyber incident, effective communication is crucial to minimize damage and confusion.
- **Implementation:** Develop and test a comprehensive crisis communication plan. Ensure all employees know the communication hierarchy and procedures during a breach.

Cultural Sensitivity in Global Organizations

- **Recognizing Diversity:** Global organizations must consider cultural differences in cybersecurity practices.
- **Implementation:** Tailor security training and communication strategies to respect and leverage cultural diversity, ensuring they resonate across different regions and teams.

Zero Trust Architecture

- **The Shift to Zero Trust:** Traditional perimeter-based security models are becoming obsolete.
- **Implementation:** Discuss the principles of Zero Trust Architecture and how adopting a Zero Trust approach can bolster an organization's cyber resilience.

Ethical Hacking and Red Teaming

- **Proactive Defense:** Ethical hackers and red teams simulate real-world attacks to identify vulnerabilities before malicious actors do.
- **Implementation:** Regularly engage ethical hackers and conduct red team exercises to uncover and address security gaps.

Legal and Regulatory Compliance

- **Staying Ahead of Regulations:** Compliance with legal and regulatory requirements is crucial for maintaining a cyber-resilient culture.
- **Implementation:** Regularly review and update security policies to ensure they align with current laws and regulations. Provide training on compliance requirements.

CONCLUSION

Building a cyber-resilient culture is an ongoing process, but the benefits are undeniable. A more secure organization translates to protected data, reduced downtime, and enhanced trust with customers and stakeholders. Take action today! Develop a customized security awareness program, conduct a security culture assessment to identify areas for improvement, and empower your employees to become your first line of defense. By investing in your cyber resilience, you're investing in the future of your organization.

CYBER CHUCKLES

- **Why did the security consultant bring a ladder to work?** To help everyone reach the next level of security awareness!
- **Why don't cybersecurity experts tell jokes about phishing?** Because they always get caught!
- **What's a hacker's favorite board game?** Risk – but they never follow the rules!
- **Why did the cyber team attend improve classes?** To prepare for any scenario!
- **How does a cybersecurity professional take their coffee?** With a strong password, of course!
- **What do you call an incident response plan that's never practiced?** A disaster waiting to happen!
- **Why did the employee bring an umbrella to the cybersecurity meeting?** In case of a data leak!
- **Why did the cybersecurity analyst stay calm during the breach?** Because panic is a hacker's best friend!

Part III

Dealing with uncertainty and complexity

Chapter 7

When the fog rolls in

Dealing with uncertainty in cybersecurity

Spoiler Alert!

- Feeling lost in the fog of cybersecurity uncertainty? Don't worry, this chapter has some tips.
- Learn how to deal with the unknown using fancy techniques like "cognitive reappraisal" and "stochastic modeling" (don't worry, they sound scarier than they are).
- This chapter even explores the idea of "embracing uncertainty" (because sometimes, you just gotta roll with the punches).

INTRODUCTION

The digital era has ushered in an unprecedented integration of technology into daily life, juxtaposing the benefits of connectivity with the vulnerabilities to cyber threats. This paper begins with a consideration of a notable cyberattack, not merely as an isolated incident but as a representation of the broader human and organizational challenges posed by such threats. This narrative foreground the exploration of "Navigating Uncertainty," a concept that extends beyond the confines of cybersecurity to encompass the realm of personal resilience.

Cyber threats, characterized by their evolving and elusive nature, pose a continuous challenge. The complexity of these threats lies not only in their technical sophistication but also in the unpredictability they introduce. Thus, preparedness transcends technological solutions, requiring psychological resilience and strategic foresight from both individuals and organizations.

This document investigates ten critical themes that strengthen cybersecurity frameworks and enhance personal resilience: cognitive reappraisal, strategic ignoring, stochastic modeling, philosophical exploration, meta-learning, experimental living, quantified self, scenario planning, and antifragility. Each theme is examined through a dual perspective: its implementation in cybersecurity measures and its significance for individual resilience, underscoring the interconnectedness of digital security and personal well-being.

Cognitive reappraisal challenges us to reinterpret cyber threats as opportunities for organizational learning and personal growth. **Strategic ignoring** emphasizes the prioritization of threats and information, advocating for focused attention on significant challenges. **Stochastic modeling** introduces probabilistic approaches to risk assessment and decision-making, relevant to both cybersecurity and personal choices. **Philosophical exploration** encourages ethical deliberation and mental equilibrium in the face of digital dilemmas.

Further, **meta-learning** and **experimental living** promote adaptability and innovation, essential qualities in the dynamic landscape of cybersecurity and personal development. The concept of the **quantified self** mirrors the data analytics employed in cybersecurity,

DOI: 10.1201/9781003562184-11

advocating for a data-informed approach to self-improvement. **Scenario planning** prepares us for diverse outcomes, enhancing resilience against unforeseen events. Lastly, **antifragility** propels us beyond mere survival of adversities, advocating for growth and strengthening through challenges.

In traversing these themes, the paper elucidates the synergistic relationship between cybersecurity strategies and personal resilience techniques. This comprehensive approach not only equips organizations to combat digital threats but also empowers individuals to navigate uncertainty with confidence and agility.

The discourse on "Navigating Uncertainty" advocates for an integrative defense strategy that marries technical safeguards with personal resilience, offering a robust response to the cyber threats of the contemporary world. It underscores the significance of human adaptability and the collective strength in addressing the challenges that lie ahead.

COGNITIVE REAPPRAISAL

Cybersecurity

Cognitive reappraisal, a concept rooted in psychological research, refers to the process of reinterpreting a situation to alter its emotional impact. In the realm of cybersecurity, this strategy can transform the perception of cyber threats from insurmountable challenges to valuable opportunities for learning and growth. For instance, threat modeling, traditionally used to identify potential security threats and vulnerabilities, can be enhanced by incorporating an understanding of cognitive biases, thereby enabling security teams to anticipate and mitigate threats more effectively.

Organizations can further leverage cognitive reappraisal through attack simulations and incident response exercises, viewing these activities not as mere preparations for potential threats but as invaluable learning opportunities. Such exercises foster a growth mindset among security personnel, encouraging them to view cybersecurity incidents as chances to strengthen systems and refine response strategies.

However, this approach requires navigating the challenges associated with overcoming negativity bias – the tendency to focus more on negative outcomes than positive ones. Cultivating a culture that emphasizes learning from security incidents, without dwelling on the negative, can enhance the resilience of cybersecurity teams.

The individual

On a personal level, cognitive reappraisal can be a powerful technique for managing the stress and anxiety associated with cyber-related threats, such as data breaches or identity theft. Individuals can apply cognitive restructuring techniques to reframe their fears about digital security, viewing potential cyber threats as manageable challenges rather than insurmountable obstacles.

Mindfulness exercises and positive self-talk can further support individuals in cultivating a calm and confident approach to navigating the digital world. These practices help in mitigating anxiety by focusing on controllable aspects of one's cybersecurity posture, such as practicing good digital hygiene and staying informed about potential threats.

Research in cognitive behavioral therapy (CBT) underscores the effectiveness of such strategies in managing anxiety disorders, offering valuable insights that can be applied to managing cyber-related stress. By adopting CBT techniques, individuals can enhance their emotional resilience, reducing the impact of cyber threats on their mental well-being.

Connecting the dots

The synergy between organizational practices of cognitive reappraisal in cybersecurity and individual strategies for managing cyber-related anxiety is evident. Both approaches lead to increased resilience: organizations become better equipped to handle and learn from cyber incidents, while individuals develop a more robust psychological defense against the stress associated with such threats.

By embracing cognitive reappraisal, both at an organizational and personal level, the cybersecurity community can foster an environment where challenges are viewed as opportunities for growth. This shift in perspective is crucial for building a resilient digital society capable of navigating the complexities of the cyber world with confidence and agility.

STRATEGIC IGNORING

Cybersecurity

In cybersecurity, strategic ignoring involves the deliberate prioritization of security alerts to concentrate resources on the most significant threats. This approach is crucial in an environment where security teams are often inundated with alerts, many of which may be false positives or of low priority. By employing risk scoring and threat intelligence filtering, organizations can identify and focus on alerts that pose a genuine risk to their infrastructure.

Automated response systems play a pivotal role in facilitating strategic ignoring by automatically handling lower-priority threats, thereby allowing human analysts to dedicate their attention to complex or high-risk issues. This balance between automation and human judgment is essential for maintaining an effective cybersecurity posture. However, this approach is not without its challenges, notably the risk of alert fatigue, where the sheer volume of alerts may desensitize analysts to threats, and the importance of ensuring that automation does not overlook nuanced or emerging threats.

The individual

For individuals, strategic ignoring can be a valuable tool in managing information overload and maintaining mental well-being in the digital age. Practical techniques such as time management strategies, selective engagement with news and social media, and controlling app notifications can help individuals focus on information that is truly relevant and beneficial to them.

The concept of attention residue, where part of one's cognitive capacity remains focused on a previous task, underscores the importance of strategic ignoring for personal productivity and mental health. Mindfulness practices can be particularly effective in training the mind to focus on the present moment, improving decision-making and reducing the stress associated with juggling multiple streams of information.

Research into mindfulness and cognitive science highlights the benefits of such practices in enhancing concentration and cognitive flexibility, providing a solid foundation for individuals to apply strategic ignoring in their personal and professional lives.

Connecting the dots

The parallels between strategic ignoring in cybersecurity and personal information management are clear. In both domains, the ability to discern what to ignore and what to focus on is crucial for resilience and effectiveness. For organizations, this means improved threat detection and response capabilities. For individuals, it leads to better mental clarity, reduced stress, and enhanced decision-making.

By applying the principles of strategic ignoring, both cybersecurity professionals and individuals can cultivate a more focused and deliberate approach to their work and lives. This strategy not only contributes to a more robust cybersecurity posture but also supports a healthier, more balanced relationship with the digital world.

STOCHASTIC MODELING

Cybersecurity

Stochastic modeling represents a sophisticated approach to understanding and mitigating cyber risks by incorporating randomness and probabilistic outcomes into the analysis. Security teams can utilize Monte Carlo simulations and other stochastic methods to estimate the likelihood of various security incidents and their potential impacts. This probabilistic approach allows for a more nuanced risk assessment, enabling organizations to allocate resources more effectively and prepare for a wide range of potential threats.

By understanding that threat actors also employ similar probabilistic models to plan their attacks, cybersecurity professionals can better anticipate and counteract these strategies. However, reliance on stochastic models also necessitates acknowledgment of their limitations. These models are based on available data, which may not fully capture emerging threats or zero-day vulnerabilities. Hence, a continuous adaptation and learning process is crucial for maintaining the relevance and effectiveness of these models.

The individual

For individuals, probabilistic thinking involves applying the principles of probability to everyday decision-making under uncertainty. This mindset encourages the consideration of multiple outcomes and their respective likelihoods, which can be particularly useful in personal risk management and financial planning. Encouraging the use of scenario planning and cost-benefit analysis can help individuals navigate personal and professional uncertainties with greater confidence.

Behavioral economics offers insights into overcoming common cognitive biases, such as overestimation of rare events or underestimation of common risks, which can skew judgment. By adopting probabilistic thinking, individuals can make more informed decisions, balancing risks, and opportunities more effectively.

Connecting the dots

Incorporating probabilistic thinking into both cybersecurity strategy and personal decision-making fosters a more informed, flexible approach to navigating uncertainty. For cybersecurity teams, this means being better prepared for a range of potential threats. For individuals, it enhances the ability to make sound decisions in the face of uncertainty.

The convergence of stochastic modeling in cybersecurity with probabilistic thinking in personal contexts underscores a shared foundation: the recognition and management of uncertainty. By embracing these approaches, both organizations and individuals can achieve a higher degree of preparedness and resilience, effectively mitigating risks in an unpredictable world.

PHILOSOPHICAL EXPLORATION

Cybersecurity

Philosophical exploration in cybersecurity involves applying timeless principles from philosophical disciplines to enhance organizational resilience against cyber threats. Concepts from Stoicism, for instance, can be particularly relevant, emphasizing the value of focusing on elements within one's control while accepting the inevitability of external threats. This mindset encourages cybersecurity teams to invest in proactive defenses and continuous improvement of internal processes, rather than expending undue energy on the unpredictable nature of cyber threats.

Epicureanism, with its focus on minimizing pain, can similarly guide cybersecurity strategies by advocating for measures that reduce the potential impact of security breaches. By adopting a philosophical lens, organizations can also navigate ethical dilemmas and decision-making in cybersecurity, ensuring actions align with broader values and societal norms.

The application of philosophical principles aids in cultivating a culture that is both ethically grounded and resilient, guiding strategic choices and fostering a balanced response to cyber incidents.

The individual

On a personal level, philosophical exploration can fortify mental resilience in the face of digital uncertainties. Adopting Stoic principles, individuals can learn to distinguish between their reactions to events and the events themselves, reducing anxiety related to cyber threats. Practices in mindfulness, acceptance, and gratitude, influenced by various philosophical teachings, offer powerful tools for managing digital stress, enhancing personal well-being.

Incorporating philosophical insights into daily life can inspire individuals to approach digital challenges with equanimity, fostering a sense of inner peace amidst the chaos of the digital age. Literature and personal philosophies that emphasize growth, adaptability, and the acceptance of change can further support individuals in navigating the complexities of modern living.

Connecting the dots

The intersection of philosophical exploration in cybersecurity and personal resilience highlights a common pursuit: achieving a measured and thoughtful approach to uncertainty. For organizations, philosophical principles provide a framework for ethical decision-making and strategic resilience. For individuals, these principles offer pathways to mental fortitude and well-being.

This shared exploration underlines the importance of grounding both cybersecurity practices and personal life choices in deeper philosophical understanding. By embracing philosophical exploration, both domains can contribute to building a more thoughtful, resilient, and ethically aware digital society.

META-LEARNING

Cybersecurity

Meta-learning, or "learning to learn," is a critical capability within cybersecurity that emphasizes the continuous acquisition and application of knowledge to improve security practices. In the dynamic landscape of cyber threats, where attackers constantly evolve their tactics, organizations must foster a culture of learning and adaptability. By implementing frameworks such as "Adversarial Thinking," cybersecurity teams can cultivate the ability to anticipate, understand, and mitigate novel threats by learning from both past incidents and emerging trends.

This approach encourages not just the acquisition of technical skills but also the development of critical thinking and problem-solving abilities. The challenges of overcoming resistance to change and the importance of building a resilient security culture underscore the need for meta-learning. Organizations that successfully embed meta-learning into their cybersecurity strategies can rapidly adapt to new threats, thereby maintaining a robust defense posture.

The individual

For individuals, meta-learning involves developing strategies for effective learning and personal growth. This encompasses cultivating a growth mindset, embracing active learning methodologies, and establishing feedback loops for continuous improvement. Resources such as online courses, self-help books, and communities of practice can support individuals in their pursuit of lifelong learning.

The benefits of meta-learning extend beyond professional development to enhance cognitive function and personal resilience. Research has shown that engaging in lifelong learning activities can mitigate cognitive decline and increase mental agility, thereby supporting individuals in navigating both personal and professional challenges with greater efficacy.

Connecting the dots

The principles of meta-learning, applied in both cybersecurity and personal development contexts, underscore the importance of adaptability and continuous improvement in navigating the complexities of the modern world. For cybersecurity teams, this means staying ahead of evolving threats through ongoing learning and adaptation. For individuals, it involves cultivating the skills and mindset necessary for lifelong learning and personal growth.

By embracing meta-learning, both organizations and individuals can build adaptable, resilient systems and selves, capable of responding to change and uncertainty with agility and confidence. This synergy between organizational learning cultures and individual development strategies contributes to a more secure and adaptable digital environment.

EXPERIMENTAL LIVING

Cybersecurity

Experimental living within the realm of cybersecurity emphasizes the value of adopting a proactive and exploratory approach to security testing and evaluation. By engaging in activities such as "Red Teaming" and ethical hacking, organizations can simulate potential cyberattacks under controlled conditions, uncovering vulnerabilities before they can be exploited by malicious actors. This methodology allows for the identification of weaknesses not just in technical systems but also in human processes and response mechanisms.

The principle of experimental living encourages a mindset of innovation and continuous improvement, where cybersecurity defenses are constantly evaluated and updated based on empirical findings. However, this approach also necessitates a careful balance between innovation and risk management, ensuring that experimentation does not inadvertently compromise security.

The individual

For individuals, experimental living involves adopting an open and curious attitude toward personal growth and problem-solving. This can mean stepping outside of one's comfort zone, trying new strategies for managing digital well-being, or adopting new technologies with a critical yet open-minded perspective. Methods such as design thinking and rapid prototyping can be applied to personal challenges, allowing individuals to creatively explore solutions and learn from the outcomes.

Experimental living fosters adaptability and innovation, qualities that are increasingly important in navigating the complexities of the digital age. It encourages individuals to view challenges as opportunities for learning and growth, rather than obstacles to be avoided.

Connecting the dots

The concept of experimental living, applied to both cybersecurity and personal development, highlights the importance of an exploratory and proactive stance toward challenges. In cybersecurity, it manifests as a commitment to continually testing and refining security measures through practical experimentation. For individuals, it represents an approach to life that values learning through experience, adaptability, and the courage to try new things.

By embracing experimental living, organizations and individuals alike can enhance their ability to respond to and thrive amidst change. This philosophy not only contributes to more robust cybersecurity practices but also supports a more adaptable and resilient approach to personal and professional life.

QUANTIFIED SELF

Cybersecurity

The concept of the quantified self, traditionally focused on tracking personal data for self-improvement, finds a parallel in cybersecurity through the meticulous analysis of security data. Organizations leverage Security Information and Event Management (SIEM) systems and other analytics tools to collect, monitor, and analyze vast amounts of security data in real time. This approach enables the identification of patterns, anomalies, and potential threats, facilitating informed decision-making and proactive threat mitigation.

By quantifying security metrics – such as incident response times, patching cadences, and the frequency of security breaches – organizations can establish benchmarks and measure the effectiveness of their cybersecurity strategies. This data-driven approach to cybersecurity not only enhances threat detection and response capabilities but also supports continuous improvement through the objective evaluation of security practices.

However, the challenges of data privacy and ethical considerations are paramount. Organizations must navigate the balance between comprehensive data collection for security purposes and the responsible handling of sensitive information.

The individual

For individuals, the quantified self-movement encourages the use of technology to track various aspects of personal life, from physical activity and diet to sleep patterns and productivity. This self-tracking can lead to greater self-awareness and actionable insights, enabling individuals to make data-informed decisions about their health, habits, and time management.

In the context of digital well-being, individuals can apply the principles of the quantified self to manage their digital consumption, monitor their online security practices, and enhance their digital literacy. Tools such as app usage trackers and digital detox apps can help individuals gain insights into their digital behaviors and implement changes to improve their online experiences and security.

Connecting the dots

The intersection of the quantified self in cybersecurity and personal development underscores the value of a data-driven approach to both organizational security and individual well-being. In cybersecurity, it facilitates a proactive and informed security posture. For individuals, it offers a pathway to personal growth and improved digital health.

By adopting the principles of the quantified self, both organizations and individuals can achieve a deeper understanding of their respective domains, leading to more effective and tailored strategies for security and personal development. This shared emphasis on data-informed decision-making highlights the broader applicability of quantification in navigating the complexities of the digital age.

SCENARIO PLANNING

Cybersecurity

Scenario planning in cybersecurity involves preparing for a wide range of potential security incidents by developing comprehensive contingency plans. This strategic approach allows organizations to anticipate and respond effectively to various scenarios, from data breaches and ransomware attacks to sophisticated state-sponsored cyber espionage. Utilizing tabletop exercises, wargaming simulations, and other scenario-based training methods, cybersecurity teams can rehearse their response to hypothetical incidents, enhancing their readiness and minimizing the potential impact of real-world attacks.

The value of scenario planning lies in its ability to foster flexibility and adaptability within security practices. By considering a broad spectrum of potential threats and outcomes, organizations can develop more resilient cybersecurity strategies that account for the unpredictable nature of cyber threats. This preparation is crucial for maintaining operational continuity and safeguarding sensitive data in the face of ever-evolving security challenges.

The individual

For individuals, scenario planning can be an effective tool for navigating personal and professional uncertainties. By preparing for different life scenarios – such as job loss, health emergencies, or financial crises – individuals can develop tailored plans that mitigate risk and reduce anxiety. Tools like personal scenario planning frameworks and emergency preparedness guides can assist individuals in envisioning possible futures and crafting strategies to navigate them successfully.

This form of planning encourages proactive thinking and decision-making, empowering individuals to take control of their futures. The psychological benefits of preparedness, including reduced stress and increased confidence, further underscore the importance of scenario planning in promoting personal resilience.

Connecting the dots

Scenario planning, whether applied to cybersecurity or personal life, emphasizes the importance of preparedness and strategic foresight in building resilience. In the realm of cybersecurity, it equips organizations with the knowledge and procedures to mitigate the impact of security incidents. For individuals, it provides a framework for anticipating and managing life's uncertainties with greater confidence.

By engaging in scenario planning, both organizations and individuals can enhance their ability to withstand and thrive in the face of unexpected events. This shared focus on strategic preparation and adaptability underscores the universal applicability of scenario planning as a tool for navigating the complexities of the modern world.

ANTIFRAGILITY AND CYBERFANTASTIC

Cybersecurity: Embracing Cyberfantastic

Antifragility in cybersecurity is a transformative approach that designs systems not just to resist attacks but to adapt and strengthen from each challenge encountered. This concept is exemplified in the "Cyberfantastic" framework, a pioneering development in cybersecurity that operationalizes the principle of antifragility. "Cyberfantastic" strategies ensure that cybersecurity infrastructures learn from attacks, dynamically adjusting to improve defense mechanisms continuously. This adaptive capacity is crucial for staying ahead of increasingly sophisticated cyber threats.

For an in-depth discussion on "Cyberfantastic" and its implementation in cybersecurity, readers are directed to the white paper available within this book.

This document provides a comprehensive overview of how antifragile principles can revolutionize cybersecurity practices, making systems more resilient and responsive to the evolving digital landscape.

The individual: Cultivating personal antifragility

For individuals, antifragility involves developing a mindset and lifestyle that thrives on challenges, stress, and change. This concept goes beyond mere resilience; it's about using adversity as a catalyst for growth, learning, and improvement. Personal antifragility can be fostered through various practices, such as:

- **Embracing Challenges:** Viewing difficulties as opportunities to learn and adapt, rather than obstacles to avoid.
- **Continuous Learning:** Engaging in lifelong learning to enhance skills and adaptability, preparing oneself for unforeseen changes and challenges.
- **Stress Inoculation:** Gradually exposing oneself to stressors in controlled amounts to build tolerance and resilience.
- **Feedback Loops:** Actively seeking and applying feedback to iterate and improve personal strategies for dealing with adversity.

By incorporating these practices, individuals can enhance their capacity to navigate life's uncertainties with confidence and agility. Cultivating personal antifragility means not just surviving in the face of adversity but thriving, growing stronger and more capable because of the challenges faced.

CONNECTING THE DOTS: INTEGRATION AND PRACTICE

The exploration of dual approaches in cybersecurity and personal resilience, through the lenses of cognitive reappraisal, strategic ignoring, stochastic modeling, philosophical exploration, meta-learning, experimental living, quantified self, scenario planning, and antifragility, culminates in a holistic strategy for navigating uncertainty. This chapter seeks to synthesize these concepts, highlighting the integration and practical application of these strategies in both organizational and individual contexts.

INTEGRATION IN CYBERSECURITY AND PERSONAL DEVELOPMENT

The intersection of cybersecurity practices with personal resilience strategies underscores a fundamental truth: the digital and personal realms are intertwined, with advancements in one area often mirroring or complementing the other. For instance, the application of meta-learning principles not only enhances organizational adaptability to cyber threats but also promotes individual growth and lifelong learning. Similarly, the practice of scenario planning prepares organizations for potential cyber incidents while helping individuals navigate personal uncertainties with confidence.

PRACTICAL APPLICATION AND SYNERGY

To effectively integrate these approaches, organizations and individuals can adopt a mindset that values continuous improvement, adaptability, and proactive engagement with challenges. For organizations, this might involve incorporating "Cyberfantastic" principles into cybersecurity training, emphasizing the importance of learning from every security incident to build antifragile systems. For individuals, it could mean applying the quantified self-approach to personal cybersecurity, using data to inform better online habits and security measures.

MOVING FORWARD: A UNIFIED APPROACH

The journey toward resilience, both digital and personal, is ongoing and iterative. It requires a commitment to applying the insights gained from each of the explored themes, continuously seeking ways to integrate these strategies into daily practices. Organizations can lead by example, fostering cultures that celebrate experimentation, learning from failures, and sharing knowledge across teams. Individuals can mirror these practices by embracing lifelong learning, cultivating flexibility, and applying a reflective and experimental approach to personal challenges.

CONCLUSIONS AND RECOMMENDATIONS

This white paper has traversed the dual landscapes of cybersecurity and personal resilience, offering a comprehensive exploration of strategies designed to navigate the complexities and uncertainties of the digital age. From cognitive reappraisal to antifragility, and the pioneering concept of "Cyberfantastic," we have uncovered the synergies between organizational security measures and individual resilience practices. Here, we distill our exploration into key conclusions and actionable recommendations for both domains.

Conclusions

1. **Interconnectedness of Cybersecurity and Personal Resilience:** The digital threats we face and the personal challenges we encounter are often reflections of the same underlying uncertainties. A holistic approach, integrating cybersecurity strategies with personal resilience techniques, is essential for comprehensive protection and growth.
2. **Dynamic Adaptation and Learning:** Both cybersecurity systems and individuals thrive on continuous learning, adaptation, and the proactive embrace of challenges. The concept of "Cyberfantastic" exemplifies this, advocating for systems that evolve to become more robust in response to attacks.
3. **Antifragility as a Goal:** Moving beyond mere survival, the aim for both organizations and individuals should be antifragility – improving in response to stressors and challenges. This requires a shift in mindset from reactive to proactive, from static defense to dynamic growth.

Recommendations

For Organizations

1. **Foster a Culture of Continuous Learning:** Encourage the adoption of meta-learning strategies to enhance adaptability and innovation in cybersecurity practices. Regular training, knowledge sharing, and post-incident reviews can cultivate a learning environment.
2. **Implement "Cyberfantastic" Principles:** Integrate antifragile concepts into cybersecurity frameworks, designing systems that not only resist but also learn from cyber threats, thereby continuously improving their defenses.
3. **Engage in Scenario Planning:** Regularly conduct scenario planning exercises to prepare for a range of potential cyber incidents, ensuring that your organization can respond quickly and effectively to any situation.

For Individuals

1. **Cultivate Personal Antifragility:** Embrace challenges as opportunities for growth. Apply principles of experimental living to personal development, continuously seeking ways to learn from adversity.
2. **Adopt a Quantified Self Approach to Digital Well-being:** Use technology to monitor and improve your digital habits, enhancing your online security posture and overall digital literacy.
3. **Practice Mindfulness and Cognitive Reappraisal:** Develop resilience against digital stressors through mindfulness practices and cognitive reappraisal, reframing the way you perceive and react to cyber threats.

MOVING FORWARD

As we advance into an increasingly interconnected and digitalized world, the lines between cybersecurity and personal resilience will continue to blur. The strategies and principles outlined in this paper provide a roadmap for navigating these challenges with confidence and agility. By integrating these approaches into our organizational cultures and personal lives, we can build a more resilient, antifragile future, ready to thrive amidst the uncertainties of the digital age.

CYBER CHUCKLES: DANCING IN THE DIGITAL RAIN OR NAVIGATING THE FOG OF CYBERSECURITY

Embarking on a journey through the foggy landscape of cybersecurity, where uncertainties loom like shadows, this chapter: "When the Fog Rolls In: Dealing with Uncertainty in Cyber Security" serves as a beacon of hope. This chapter doesn't just illuminate the path; it equips you with a high-powered flashlight and a compass, metaphorically speaking, to navigate through the mist with confidence.

Hacking through the haze: A cybersecurity odyssey

Imagine cybersecurity as an ever-evolving labyrinth, where each turn could lead to new challenges or hidden opportunities. As we delve into the realm of unknowns, fancy techniques like "cognitive reappraisal" transform our perception of cyber threats from daunting specters to puzzles waiting to be solved. "Stochastic modeling" morphs from an arcane spell into a strategic tool, mapping out the probabilities of cyber threats with the precision of a cartographer charting unknown waters.

As the narrative unfolds, we're introduced to the art of "embracing uncertainty." It's akin to learning to dance in the rain, where each droplet of doubt waters the seeds of resilience and growth. This chapter is not just about surviving the storm but thriving within it, turning unpredictability into a strategic advantage.

Tales of triumph: The Cyberfantastic voyage

Woven into this journey is a tale that resonates with the spirit of every underdog story ever told. It's the saga of a modern IT security company, navigating through the tempest of digital threats. Their odyssey is marked by trials and triumphs, a testament to the power of collective resilience and the strength found in shared responsibility.

Picture a team that once stood on the brink of digital defeat, their defenses besieged by the relentless onslaught of cyber threats. Yet, instead of succumbing to despair, they chose to rewrite their fate. Armed with tools of cognitive reappraisal, they learned to see each challenge as a lesson, each failure as a steppingstone.

Stochastic modeling became their navigation chart, guiding them through the turbulent seas of cyber risks with foresight and agility. And in their most daring move, they embraced uncertainty, turning it into their shield and sword. This bold strategy allowed them to anticipate the unpredictable, adapt to the unforeseen, and emerge stronger from each encounter.

As the fog of uncertainty began to lift, what emerged was not just a company that had weathered the storm but one that had mastered the art of sailing through it. Their story is a beacon of hope for all navigating the complex seas of cybersecurity, a reminder that with the right mindset and tools, even the most uncertain journeys can lead to triumphant shores.

Epilogue: A lighthouse in the fog

This chapter doesn't just guide you through the fog; it teaches you to chart your own course through it. It's a testament to the power of adaptability, the strength of a united team, and the resilience of the human spirit in the face of digital uncertainty.

As we close this chapter, we're reminded that in the ever-changing landscape of cybersecurity, uncertainty isn't a curse but an opportunity. It's a chance to grow, to learn, and to emerge not just unscathed but stronger, wiser, and more prepared for whatever lies ahead.

In this cyber odyssey, the fog doesn't obscure our path; it illuminates the resilience, ingenuity, and spirit of those willing to navigate through it. So, as we move forward, let's embrace the fog, for within it lies the key to mastering the art of cybersecurity and beyond.

Chapter 8

Dynaxity and beyond

A model for the modern cyber landscape

Spoiler Alert!

- Imagine a cybersecurity landscape with zones – static, dynamic, turbulent, and chaotic (think tea party to mosh pit).
- This chapter explains how to adapt your security strategy to each zone (because a one-size-fits-all approach just won't work).
- Get ready for some fun scenarios: a global corporation fighting cyberattacks, a hospital fending off digital disease, and a store dealing with online shopping sprees gone rogue.

INTRODUCTION

The landscape of information security is fraught with challenges that are both dynamic and complex. The term "Dynaxity," a blend of dynamics and complexity, emerges as a crucial conceptual framework for navigating this landscape. It provides a lens through which organizations can understand the interplay between the ever-changing nature of cyber threats and the intricate environments they operate within. This introduction sets the stage for an in-depth exploration of Dynaxity, underscoring its significance in the realm of information security.

Dynaxity, as a framework, assists organizations in categorizing and responding to challenges based on the level of dynamism and complexity they present. In a field where threats evolve with alarming speed and the digital infrastructure becomes increasingly convoluted, grasping the principles of Dynaxity is more than an academic exercise – it's a strategic imperative. By applying this model, organizations can better anticipate, prepare for, and mitigate the risks associated with information security in a manner that is both proactive and informed.

As we delve into this chapter, we aim to elucidate the essence of Dynaxity and its application to information security. We will explore the model's four distinct zones – static, dynamic, turbulent, and chaotic – each representing a different mix of dynamics and complexity. This exploration will reveal how understanding one's position within these zones can guide the formulation and implementation of effective information security strategies. Through this lens, the relevance of Dynaxity to information security becomes clear, offering a novel approach to safeguarding digital assets in an increasingly uncertain cyber landscape.

The introduction of Dynaxity into the discourse on information security heralds a shift in how organizations conceptualize and tackle digital threats. It challenges traditional models that may not fully account for the fluidity and intricacy of modern cyber

DOI: 10.1201/9781003562184-12

challenges. As we proceed, this chapter will unfold the layers of Dynaxity, demonstrating its pivotal role in crafting resilient, responsive, and nuanced information security practices.

DYNAXITY EXPLAINED

Diving into the concept of Dynaxity, it's imperative to understand its foundation: the dynamic interplay between dynamics (the rate of change within an environment) and complexity (the degree of interconnected components and variables). This section elucidates the four zones of Dynaxity – static, dynamic, turbulent, and chaotic – and contextualizes them within the sphere of information security, offering a granular view of the challenges and scenarios unique to each zone.

Static zone

This environment is characterized by minimal change and complexity. Threats are well understood and relatively stable, making traditional, well-established security measures and protocols effective. For instance, an organization operating in a static zone might face routine malware threats that can be countered with standard antivirus software and firewall protections.

Dynamic zone

As either complexity or dynamics begin to increase, the need for more flexible and adaptive security strategies becomes apparent. In this zone, new technologies or expanding digital footprints might introduce novel vulnerabilities, requiring continuous monitoring and timely updates to security measures. An example could include adapting to the security challenges posed by the adoption of cloud services, where the dynamic allocation of resources necessitates a more vigilant approach to access control and data encryption.

Turbulent zone

High levels of both dynamics and complexity define this environment. Organizations in the turbulent zone face unpredictable threats that demand advanced analytical capabilities and adaptive security strategies. For instance, dealing with sophisticated cyber-espionage campaigns targeting proprietary data would fall into this category, requiring a mix of real-time threat intelligence, anomaly detection, and incident response agility.

Chaotic zone

The apex of challenge, is characterized by extreme dynamics and complexity. Traditional approaches to security may be inadequate, necessitating innovative and highly adaptive measures. A scenario fitting this zone could involve responding to a zero-day exploit within a critical infrastructure system, where the lack of prior knowledge and the high stakes demand swift, creative problem-solving and emergency response measures.

Understanding these zones and the specific challenges they present is crucial for developing effective information security strategies. Each zone demands a tailored approach, leveraging different tools, methodologies, and mindsets to navigate the inherent risks successfully.

This exploration of the Dynaxity zones lays the groundwork for discussing how information security policies, standards, guidelines, and procedures can be adapted to effectively address the diverse challenges presented across these environments.

POLICY FRAMEWORK IN INFOSEC

The policy framework within information security governance plays a foundational role in establishing, guiding, and enforcing the measures that organizations take to protect their digital assets. This framework is structured hierarchically, comprising policies, standards, guidelines, and procedures. Each layer serves a distinct purpose and, when effectively aligned, creates a robust governance model that can adapt to the varying degrees of dynamics and complexity outlined in the Dynaxity model (Figure 8.1).

The hierarchy explained:

- **Policies:** These are high-level statements that set the overall direction and principles for information security within an organization. Policies are typically broad, outlining the goals and expectations for security without delving into specifics. They are the cornerstone of an infosec governance framework, reflecting the organization's commitment to securing its assets.
- **Standards:** Standards provide a more detailed and specific set of requirements that must be met to comply with the policies. They are often technical in nature, offering clear criteria for technology configurations, security controls, and behaviors that support policy objectives. Standards ensure consistency and reliability in how security measures are implemented and maintained.
- **Guidelines:** While policies and standards are generally mandatory, guidelines offer recommended practices that can be adapted as needed. They provide advice on how to meet the standards in a way that accommodates different situations or technologies, offering flexibility within the framework's structure.
- **Procedures:** The most detailed component of the framework and procedures provide step-by-step instructions for implementing the policies, standards, and guidelines. They are actionable and specific, ensuring that day-to-day operations align with the overarching security goals.

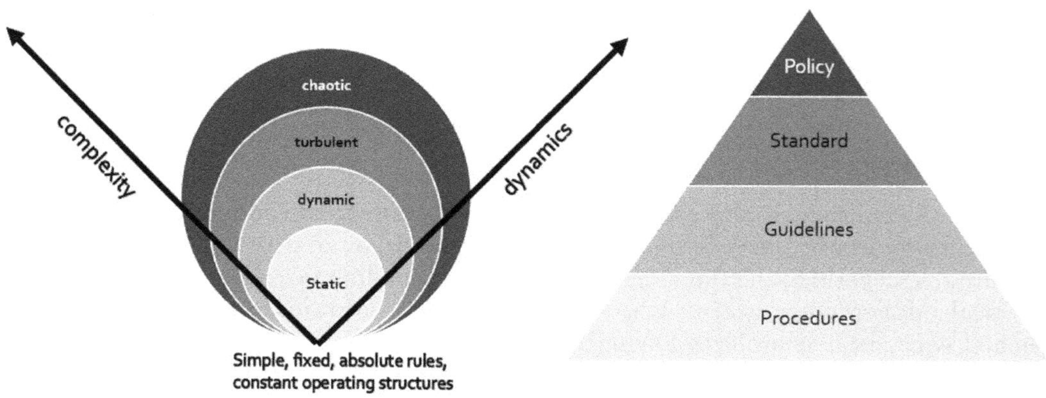

Figure 8.1 Dynaxity model (compound of dynamics and complexity) applied to the Field of Information Security, especially Policy development.

ADAPTATION ACCORDING TO DYNAXITY ZONES

An organization's positioning within the Dynaxity model significantly influences how these elements of the policy framework are emphasized and applied. The dynamic and complex nature of the cybersecurity landscape requires that policies, standards, guidelines, and procedures be flexible and responsive to the environment's characteristics.

- In **static zones**, where threats and technologies change minimally, policies and standards might be more prescriptive, offering clear, unchanging directives that ensure stability and compliance.
- Moving into **dynamic zones**, the importance of guidelines increases, providing the necessary flexibility to adapt to new technologies or emerging threats without the need for constant policy or standard revisions.
- In **turbulent and chaotic zones**, procedures become crucial, needing to be both detailed for immediate action and adaptable for unforeseen challenges. The rapid response to incidents and the ability to innovate in the face of novel threats are paramount, requiring a governance model that supports quick decision-making and flexibility.

This section has outlined the critical components of an information security policy framework and how they can be dynamically adjusted according to an organization's position within the Dynaxity model. Understanding and applying this adaptive approach is essential for developing effective governance that not only protects against current threats but is also resilient and responsive to future challenges.

Next, we'll discuss the application of Dynaxity to information security, exploring how recognizing an organization's position within the Dynaxity model can inform and guide its security efforts, tailoring strategies to effectively navigate the complexities of the cyber landscape.

APPLICATION OF DYNAXITY TO INFOSEC

Understanding an organization's position within the Dynaxity model is not just an academic exercise; it's a strategic imperative that can significantly influence the focus and governance of its information security efforts. Recognizing whether an organization operates within a static, dynamic, turbulent, or chaotic environment allows for the tailoring of security strategies to not only address current challenges but also anticipate future threats. This part of the chapter explores how the Dynaxity model can be applied to guide information security practices effectively.

Strategic tailoring for each zone

- **Static Environments:** In these relatively stable and predictable settings, information security strategies can afford to be more standardized and procedural. Emphasizing comprehensive risk assessments and regular audits ensures that the established security measures remain effective against known threats. The focus here is on maintaining vigilance and reinforcing existing defenses.
- **Dynamic Environments:** As the level of complexity or dynamics increases, information security strategies must become more flexible and responsive. This might involve implementing advanced threat detection systems that leverage machine learning to

adapt to new attack patterns or adopting cloud security practices that offer scalability and responsiveness to evolving technological landscapes.
- **Turbulent Environments:** Here, the unpredictability and complexity of threats require an information security approach that emphasizes agility and the ability to pivot quickly. Strategies might include the establishment of a dedicated incident response team capable of dealing with a wide range of scenarios and adopting a zero-trust architecture to minimize the impact of breaches.
- **Chaotic Environments:** In the most challenging settings, where threats are both highly dynamic and complex, information security strategies need to prioritize innovation and rapid adaptation. This could involve engaging in red team-blue team exercises to continually test and refine defenses, or the integration of cutting-edge technologies like artificial intelligence for real-time threat analysis and response.

Guiding focus and governance

Understanding an organization's Dynaxity zone provides invaluable insights that can shape the governance of its information security efforts. For instance, in more stable zones, governance might focus on compliance and consistency, ensuring that policies and standards are rigorously applied. As environments become more turbulent, governance needs to shift toward enabling rapid decision-making and flexibility, perhaps decentralizing certain security functions to allow for quicker responses to threats.

LEVERAGING DYNAXITY FOR ADAPTIVE SECURITY

The application of Dynaxity to information security emphasizes the need for security measures that are not only robust but also adaptable. It advocates for a security posture that can evolve in response to the changing dynamics and complexity of the cyber landscape. Key to this is the development of a security culture that values continuous learning, agility, and collaboration across all levels of the organization.

- **Continuous Learning and Improvement:** In alignment with the principle of continuous improvement inherent in Dynaxity, organizations should foster a culture of ongoing education and adaptation within their security teams. Encouraging security professionals to stay abreast of emerging threats and innovative defense mechanisms ensures the organization's security posture remains resilient.
- **Collaboration across Disciplines:** The interdisciplinary nature of Dynaxity underscores the importance of collaboration between different expertise areas within an organization. Integrating insights from data science, behavioral psychology, and business strategy can enrich information security practices, making them more comprehensive and effective.

The application of the Dynaxity model to information security offers a strategic framework for navigating the complexities of the cyber landscape. By understanding their position within the Dynaxity zones, organizations can tailor their security strategies to be as dynamic and complex as the threats they face, ensuring their defenses remain effective and adaptable in the face of uncertainty.

In the next section, we will explore hypothetical scenarios that illustrate how organizations might apply the Dynaxity model to manage information security across different levels, providing concrete examples of the model's practical implications.

HYPOTHETICAL SCENARIOS

Exploring hypothetical scenarios offers a practical lens through which to view the application of the Dynaxity model in managing information security. These scenarios or brief case studies envision how organizations at various operational levels – such as corporate, country, or branch –might leverage Dynaxity principles to address their unique information security challenges.

Scenario 1: A global corporation adapting to dynamic threats

- **Context:** A multinational corporation operates in the dynamic zone, facing challenges from rapid technological advancements and emerging cyber threats targeting its diverse global operations.
- **Application of Dynaxity:** Recognizing its position, the corporation adopts flexible cloud security solutions to accommodate fluctuating demand and introduces an AI-driven threat intelligence system that evolves with new threat patterns. Governance structures are adapted to allow regional teams to make swift security decisions in response to local threats, emphasizing a decentralized approach for agility.
- **Outcome:** Enhanced responsiveness to emerging threats across its global footprint, with improved incident detection times and a robust ability to adapt to the technological landscape's dynamics.

Scenario 2: A healthcare provider navigating turbulent times

- **Context:** Facing a surge in targeted ransomware attacks amid a global health crisis, a healthcare provider finds itself in a turbulent zone, where the complexity and unpredictability of threats are compounded by the critical nature of its digital assets.
- **Application of Dynaxity:** The provider escalates its incident response capabilities by forming a cross-functional crisis management team, including IT, legal, and communications experts, to address multifaceted threats rapidly. It also implements zero-trust network access to secure patient data further and deploys advanced analytics to predict and prevent potential breaches.
- **Outcome:** The organization manages to mitigate the impact of ransomware attacks effectively, safeguarding patient data and ensuring the continuity of critical healthcare services during a period of heightened vulnerability.

Scenario 3: A retail chain responding to chaotic disruptions

- **Context:** A retail chain experiences chaotic disruptions due to a sophisticated cyberattack that coincides with a major product launch, highlighting vulnerabilities in its e-commerce platform and physical point-of-sale systems.
- **Application of Dynaxity:** In response, the chain launches an emergency task force to manage the crisis, incorporating rapid response protocols and real-time communication channels to coordinate efforts across departments. It employs cutting-edge cybersecurity solutions, including behavioral analytics and automated incident response systems, to address the attack's complexity and dynamics.
- **Outcome:** The retail chain successfully navigates the crisis, minimizing financial loss and restoring consumer trust. Post-incident, it revises its information security strategy to include regular stress-testing of its systems against chaotic scenario simulations, strengthening its preparedness for future disruptions.

These hypothetical scenarios illustrate the practical application of the Dynaxity model across various operational levels and environments. By tailoring their information security practices according to their specific Dynaxity zone, organizations can not only address current threats more effectively but also enhance their resilience and adaptability in the face of future challenges.

The concluding section will summarize the significance of the Dynaxity model in enhancing information security governance and adaptability across different environmental contexts, emphasizing the model's value in navigating the complex and dynamic landscape of cybersecurity threats.

CONCLUSION

The exploration of the Dynaxity model within the realm of information security culminates in a profound understanding of how dynamics and complexity influence the effectiveness of cybersecurity measures. This chapter has articulated the significance of recognizing an organization's position within the Dynaxity framework, a recognition that is instrumental in tailoring information security governance and strategies to navigate the intricate landscape of cyber threats successfully.

The Dynaxity model, with its delineation into static, dynamic, turbulent, and chaotic zones, provides a nuanced lens through which organizations can assess their vulnerability to cyber threats and the requisite agility of their response mechanisms. The hypothetical scenarios presented underscore the model's practical applicability, demonstrating how organizations can adapt their information security practices to the unique challenges posed by their operational environment.

Key takeaways

- **Adaptive Security Practices:** The Dynaxity model advocates for an adaptive approach to information security, urging organizations to evolve their strategies in alignment with the level of complexity and dynamics they face. This adaptive approach ensures that cybersecurity measures remain robust and responsive to an ever-changing threat landscape.
- **Strategic Governance:** Understanding one's position within the Dynaxity zones can significantly influence information security governance, directing the focus toward either stabilizing measures in static environments or innovative, flexible strategies in more chaotic contexts. This strategic governance is crucial for maintaining an effective defense against cyber threats.
- **Continuous Improvement and Collaboration:** The model reinforces the importance of continuous improvement and cross-disciplinary collaboration in enhancing cybersecurity resilience. By fostering a culture of ongoing learning and leveraging insights from various fields, organizations can better anticipate and mitigate the impacts of cyber threats.

Looking forward

As we look to the future, the Dynaxity model's relevance to information security is poised to grow. The digital world's inherent complexity and the dynamic nature of cyber threats necessitate frameworks that can accommodate rapid changes and multifaceted challenges. Organizations that embrace the principles of Dynaxity, adapting their information security practices accordingly, will find themselves better equipped to navigate the uncertainties of the cyber landscape.

In essence, the Dynaxity model offers a foundational strategy for enhancing information security governance and adaptability across diverse environmental contexts. It empowers organizations to build layered defenses that flex with the threat, ensuring that cybersecurity measures are not only effective in the present but are also sustainable and resilient in the face of future challenges. Embracing Dynaxity is not just about responding to the complexities of today; it's about preparing for the dynamics of tomorrow, setting a course for a secure and agile digital future.

This exploration of Dynaxity within the context of information security closes with a call to action for organizations to critically assess their position within the Dynaxity framework and to strategically align their cybersecurity efforts with the principles outlined. In doing so, they can achieve a more adaptive, responsive, and ultimately effective approach to safeguarding their digital assets against the myriad threats that define our modern cyber landscape.

CYBER CHUCKLES: KEY TAKEAWAYS WITH A TWIST

Welcome to the wild world of Dynaxity, where the cyber landscape ranges from the calmness of a tea party in the static zone to the frenzy of a mosh pit in the chaotic zone. Grab your cybersecurity gear; we're going on an adventure across the zones, finding humor and strategy in equal measure.

Static zone: The tea party

Here, in the serene and orderly static zone, cybersecurity threats move about as fast as a sloth on a lazy Sunday. Strategies here are like your grandma's best tea set – reliable, time-tested, and a little old-fashioned. Perfect for a zone where the biggest excitement might be a phishing email that actually spells "phishing" correctly.

Dynamic zone: The office on wheels

Next up, the dynamic zone – it's like conducting a board meeting on a bus that's always turning corners. You're adapting, moving, and possibly getting a little motion sickness from the constant change. Here, cloud security is your best friend, and your strategy is about as flexible as a Cirque du Soleil performer, ready to bend but not break under the pressure of new technologies.

Turbulent zone: The wild west

Welcome to the turbulent zone, where cybersecurity is a rodeo, and you're the cowboy trying to stay on the bull. It's wild, unpredictable, and you'll need more than just your hat to protect you. This zone calls for a zero-trust architecture because, in this rodeo, even the clowns might be out to get you. Your incident response team is your lasso, ready to wrangle whatever cyber threat comes your way.

Chaotic zone: The mosh pit of mayhem

And now, the chaotic zone, where cybersecurity is less about strategy and more about survival. Imagine diving into a mosh pit, but everyone's a hacker, and the music is a constant barrage of zero-day exploits. Here, innovation is your armor, and quick reflexes are your best defense. It's about finding order in chaos, like trying to organize a dance routine in the middle of that mosh pit.

Adapting your cybersecurity waltz

Across all these zones, the key is to dance to the rhythm of Dynaxity – sometimes it's a waltz, sometimes a tango, and occasionally, it's just trying not to step on too many toes. It's about knowing when to lead, when to follow, and when to throw out the rulebook and just jam.

In conclusion: party favors

As we wrap up our journey through the Dynaxity model, remember: cybersecurity isn't a one-size-fits-all party dress. It's a wardrobe tailored to the occasion, from the tea party to the mosh pit. So, whether you're sipping tea or crowd surfing, keep your cybersecurity adaptable, responsive, and, most importantly, equipped with a sense of humor. Because in the digital wild west of Dynaxity, a good laugh might just be your best defense.

Chapter 9

Black Swans, Grey Rhinos, and your bottom line

Adapting financial models to cyber risk

Spoiler Alert!

- This chapter teaches you to speak the language of finance guys (because apparently, cyber risk is money now).
- Get ready for fancy terms like "Black Swan" events (unexpected cyberattacks) and "Grey Rhinos" (highly probable threats we choose to ignore).
- Learn how to build a financial model that factors in cyber risk (so you can finally show your boss why you need that extra firewall).

INTRODUCTION

The innovative endeavor of adapting the Black-Scholes-Merton (BSM) model to cybersecurity risk management signifies a pioneering leap toward integrating sophisticated financial risk management methodologies with the dynamic field of cybersecurity. This chapter sets the stage for a detailed exploration of how the foundational principles of the BSM model, traditionally employed for pricing European options in financial markets, can be reinterpreted and applied to the evaluation, optimization, and strategic management of cybersecurity investments and initiatives.

The BSM model's potential to quantify the elusive variables of cyber risk presents an enticing opportunity for cybersecurity professionals seeking to refine their risk management strategies with data-driven insights. Through a structured, multi-step process, this chapter will guide the reader through the conceptual framework development, quantitative model building, model validation and calibration, implementation, and the continuous improvement phase, emphasizing practicality and real-world application.

STEP 1: CONCEPTUAL FRAMEWORK DEVELOPMENT

Forming an interdisciplinary team

The journey begins with the assembly of an interdisciplinary team, bringing together experts from finance, cybersecurity, data science, and game theory. This collaborative approach ensures a well-rounded adaptation of BSM concepts, allowing for a rich integration of perspectives that are crucial for tackling the multifaceted challenges of cybersecurity risk.

Defining cybersecurity equivalents

A critical step in the adaptation process involves translating key BSM components into cybersecurity equivalents. For instance, "volatility" in financial terms becomes a measure of the

DOI: 10.1201/9781003562184-13

unpredictability and variability inherent in cyber threats. This step requires a deep dive into both fields to identify analogous elements and redefine them within the cybersecurity context.

Scope and attacker behavior

Determining the model's focus involves specifying the types of cyberattacks, the industries most affected, and incorporating evolving attacker strategies into the model. This ensures that the adapted framework remains relevant and effectively captures the dynamic nature of cyber threats.

STEP 2: QUANTITATIVE MODEL BUILDING

Data collection and analysis

At the heart of model adaptation is the collection and analysis of cyber incident data. This foundational step involves gathering detailed information on the frequency and types of cyberattacks, as well as assessing the current effectiveness of security measures in place. This data informs the development of the model, grounding it in empirical evidence.

Model formulation

With a comprehensive dataset in hand, the next step is to develop mathematical models that adapt BSM equations to cybersecurity. This involves defining success metrics, such as incident reduction rates or improvement in response times, to gauge the model's effectiveness and applicability in real-world scenarios.

STEP 3: MODEL VALIDATION AND CALIBRATION

Backtesting and calibration

Testing the model against historical cyber incident data is crucial for assessing its accuracy and relevance. Calibration, informed by outcomes and expert feedback, refines the model, enhancing its predictive power and ensuring it aligns with the realities of the cybersecurity landscape.

Sensitivity analysis

Performing sensitivity analysis sheds light on how variations in input parameters impact the model's outputs. This step underscores the importance of considering data limitations and expert judgment in calibrating the model, ensuring it remains robust and reliable.

STEP 4: IMPLEMENTATION AND CONTINUOUS IMPROVEMENT

Integration into risk management frameworks

Seamlessly integrating the adapted model within existing cybersecurity risk management frameworks is pivotal. Developing user-friendly interfaces ensures broader accessibility and usability, facilitating the model's adoption across different organizational levels.

Continuous data integration

To keep pace with the rapidly evolving cybersecurity landscape, mechanisms for ongoing data integration and regular model updates are established. This iterative refinement process ensures the model stays current, reflecting the latest threat intelligence and security trends.

STEP 5: SHARING KNOWLEDGE AND BEST PRACTICES

Knowledge dissemination

Documenting the model development process, conducting educational workshops, and fostering collaborations with academic and research institutions are essential for disseminating knowledge and encouraging the adoption of BSM concepts in cybersecurity risk management.

HIGHLIGHTING SUCCESSES

Presenting case studies of successful model implementations offers tangible evidence of its value, promoting wider acceptance and illustrating the practical benefits of integrating BSM principles into cybersecurity strategies.

IN MEDIAS RES

Integrating Black-Scholes-Merton concepts into cybersecurity risk management

The potential to adapt the BSM equation for cyber risk management is a compelling innovation. This approach seeks to apply the underlying principles of the BSM model, originally designed for pricing European options in financial markets, to the evaluation and optimization of cybersecurity investments and strategies.

ORIGINAL BSM EQUATION OVERVIEW

The BSM model is fundamentally expressed through the equation (Figure 9.1).

$$\frac{\partial V}{\partial t} + \frac{1}{2}\sigma^2 S^2 \frac{\partial^2 V}{\partial S^2} + rS\frac{\partial V}{\partial S} - rV = 0$$

Figure 9.1 The original Black–Scholes equation, also called the Black–Scholes–Merton equation, is a partial differential equation governing the price evolution of derivatives.

where V represents the option price, S is the price of the underlying asset, σ signifies the volatility of the asset price, r is the risk-free interest rate, and t denotes time.

CONCEPTUAL FRAMEWORK FOR CYBERSECURITY

Adapting the BSM equation to cybersecurity involves reinterpreting its components to fit the cyber risk management context:

- V **(Option Price):** Transforms into the value or cost associated with implementing cybersecurity measures.
- S **(Underlying Asset Price):** Becomes the value of the digital assets or data protected.

- σ **(Volatility):** Reflects the uncertainty or variability in the frequency and severity of cyber threats.
- r **(Risk-free Interest Rate):** Analogous to the expected return on investment (ROI) of the most secure, risk-free cybersecurity measure.
- t **(Time):** Remains as the period over which cybersecurity measures are expected to be effective.

ADAPTED BSM EQUATION FOR CYBERSECURITY

The adaptation leads to a new conceptual equation:

$$\frac{\partial C}{\partial t} + \frac{1}{2}\theta^2 A^2 \frac{\partial^2 C}{\partial A^2} + bA\frac{\partial C}{\partial A} - bC = 0$$

Figure 9.2 Adapted Black–Scholes–Merton equation for using as a calculation for cybersecurity in the dynamics between attackers and defenders.

Here, C indicates the cost/value of cybersecurity measures, A is the value of the assets being protected, θ represents the threat volatility, and b signifies the baseline security level or ROI of risk-free measures (Figure 9.2).

Implementation and Application

1. **Data Collection and Model Calibration:** Initiates with gathering data on cyber incidents to calibrate the model's parameters, focusing on estimating threat volatility (θ), asset value (A), and baseline security ROI (b).
2. **Quantitative Analysis:** Utilizes the adapted model to assess various cybersecurity measures, aiming to identify the optimal investment that minimizes risk cost-effectively.
3. **Risk Mitigation Strategy Development:** Formulates strategies prioritizing cybersecurity investments that yield the highest protection ROI, guided by the model's insights.

Challenges

- Accurately estimating model parameters, especially cyber threat volatility (θ), is complex due to the ever-evolving nature of cybersecurity threats.
- The model's sophistication necessitates advanced statistical and mathematical methodologies for effective application and interpretation.
- Continuous updates are essential to reflect the latest cyber threat landscape and ensure the model remains relevant and effective.

Integrating BSM concepts into cybersecurity offers a novel pathway to quantifying and strategically managing cyber risks. This conceptual adaptation opens doors to developing more sophisticated, data-driven approaches for optimizing cybersecurity investments, marking a significant leap forward in how organizations protect their digital assets against evolving threats.

CONCLUSION

Adapting the BSM model to cybersecurity risk management opens up new avenues for quantifying and strategically managing digital threats. This chapter has outlined a structured approach to this adaptation, offering insights into developing a sophisticated, data-driven framework for optimizing cybersecurity investments. As organizations navigate the complexities of the digital threat landscape, the integration of BSM concepts into cybersecurity strategies promises not only to enhance risk management practices but also to catalyze innovation in the field.

CYBER CHUCKLES: NAVIGATING THE FINANCIAL FRONTIER OF CYBERSECURITY

Welcome to the world where cybersecurity meets Wall Street, and your firewalls need to speak dollar bills. In this thrilling chapter, we embark on an adventure to decode the lingo of finance gurus and navigate the treacherous waters of cyber risk without losing our shirts – or our data.

The Swans and Rhinos of cyberspace

Imagine your cybersecurity landscape as a wildlife reserve, where Black Swans are those rare, unpredictable cyberattacks that leave you gasping, and Grey Rhinos are those big, obvious threats charging right at you – yet, somehow, you're too busy updating your social media status to notice.

Translating the financial Gibberish

We're taking the BSM model – a fancy formula used by financial wizards to predict stock options – and tweaking it to predict how much that extra firewall is really worth. It's like translating Shakespeare into emojis, ensuring the finance department finally gets why cybersecurity is the real MVP.

The Five-Step Dance to Cyberfinancial Genius

1. **Forming the Avengers of Finance and Cybersecurity:** First, we assemble a dream team – an interdisciplinary squad where finance nerds and cyber geeks unite. Their mission: to make the BSM model the next cybersecurity superhero.
2. **Crafting the Codebook:** Next, we dive into translating the financial mumbo-jumbo into cybersecurity speak. Volatility isn't just for stocks anymore; it's about the unpredictable world of cyber threats. It's like converting a financial recipe into a cybersecurity cookbook.
3. **The Great Data Hunt:** Armed with calculators and code, our team gathers cyber incident data like squirrels preparing for winter. This step is crucial for feeding the hungry model with all the juicy details it needs to make predictions.
4. **Unleashing the Math Wizards:** Now, the magic happens. Our team adapts the model to predict cyber risk, turning financial equations into cyber defense strategies. It's like using a crystal ball, but with more numbers and less hocus pocus.
5. **From Theory to Action:** Finally, we put our model to the test, integrating it into real-world cyber strategies. It's where the rubber meets the road, or rather, where the algorithm meets the firewall.

The conclusion: Where finance and cybersecurity live happily ever after

In the end, we've not only bridged the gap between cybersecurity and finance but also shown how preparing for Black Swans and charging at Grey Rhinos can save the day – and the bottom line. Now, cybersecurity isn't just about protecting data; it's about smart investment and proving that, yes, that extra firewall is indeed worth every penny.

So, here's to the brave souls making cyber risk a part of the financial conversation, one BSM model at a time. May your ledgers be balanced, your data secure, and your financial forecasts always sunny, with a slight chance of Black Swans.

Part IV

Advanced topics and technologies

Chapter 10

Deception and misdirection in cybersecurity

Spoiler Alert!

- **Shifting the Economic Burden:** Learn how deception techniques can make cyberattacks more costly and less attractive for attackers.
- **Advanced Deception Tactics:** Explore dynamic deception, social engineering countermeasures, and multilayered deception strategies.
- **Deception in the Supply Chain:** Understand how to implement deception in third-party systems to secure the entire supply chain.

INTRODUCTION

Imagine a high-stakes game of cat and mouse. In this case, the cat is your company's security team, desperately trying to stay ahead of cunning attackers, the mice. Traditional defenses like firewalls and intrusion detection systems (IDS) act as tripwires, alerting you when someone tries to break in. But what if you could set a trap, not just detect, but also learn about the attackers and their tactics? Enter the world of deception and misdirection in cybersecurity.

"Cybersecurity is no longer just about building walls; it's about creating a maze for attackers," says John Strand, a cybersecurity expert. This quote underscores the shift from traditional defense to a more proactive and deceptive strategy.

Deception is a powerful, yet nuanced, security strategy that goes beyond simply blocking attackers. It's about creating illusions, planting false information, and setting up honey traps to lure attackers in and gather valuable intelligence. More importantly, deception techniques can significantly alter the economic equation of cyberattacks. Traditionally, attackers have the upper hand, needing to find just one vulnerability to succeed, while defenders must protect all possible entry points. Deception flips this script by changing the economics of cyberattacks.

By deploying honeypots, honeynets, and other deceptive tactics, organizations can force attackers to waste considerable time, effort, and resources on fake systems. These deceptive measures act as time sinks, leading attackers down fruitless paths and away from valuable targets. This increased cost and effort make cybercrime less attractive and riskier. Attackers must invest more resources without the guarantee of a successful breach, thereby raising the economic stakes for them.

Furthermore, deception techniques provide defenders with the upper hand in gathering intelligence. By studying how attackers interact with these decoys, security teams can gain insights into their methods, tools, and strategies. This information is invaluable for strengthening real defenses and preparing for future attacks.

This chapter will delve into the world of deception techniques, explore their effectiveness, and discuss the ethical considerations involved in their responsible implementation. We'll

DOI: 10.1201/9781003562184-15

also examine how deception can disrupt the economic incentives of attackers, turning the tables in favor of defenders. By shifting the economic burden and extending the attack cycle, deception not only protects critical assets but also makes cybercrime a less viable and attractive endeavor for attackers.

SHIFTING FOCUS IN DECEPTION

Deception for Defense Beyond Networks

- **Cloud Platforms:** Create deceptive environments in cloud platforms to identify attackers targeting cloud infrastructure. This involves deploying fake cloud resources like storage buckets and virtual machines designed to attract and monitor malicious activity.
- **IoT Devices:** Develop deceptive Internet of Things (IoT) environments where fake IoT devices mimic real ones, luring attackers who target smart home devices, industrial control systems, or healthcare devices.
- **Social Media Profiles:** Create deceptive social media profiles that appear genuine to gather intelligence on attackers attempting social engineering or phishing attacks.

Beyond traditional networks, deception techniques can also be effectively applied to cloud platforms and IoT devices, expanding the reach of security strategies.

Deception as Misdirection, Not Just Luring

- **Fake Information Trails:** Instead of just luring attackers into honeypots, create fake data trails that lead attackers away from valuable assets. This can involve planting fake documents, credentials, or system configurations designed to misdirect.
- **Behavior Manipulation:** Use deception to influence attacker behavior, making them waste time and resources. This can be achieved by providing false clues and manipulating their decision-making processes.

Deception as a Psychological Tool

- **Exploiting Attacker Biases:** Incorporate psychological tactics into deception technology to exploit common cognitive biases of attackers. For example, creating scenarios that play on overconfidence or anchoring biases can make deceptions more believable.
- **Decision-Making Flaws:** Design deceptions that exploit known flaws in attacker decision-making processes, making them more likely to fall for traps and reveal their tactics.

EVOLVING TACTICS IN DECEPTION

Deception with AI

- **Adaptive Environments:** Utilize artificial intelligence (AI) to dynamically generate and adapt deceptive environments in real time based on attacker behavior. This creates a more complex and unpredictable scenario for attackers, making it harder for them to distinguish between real and fake systems.

- **Behavioral Analysis:** AI can analyze attacker behavior patterns and generate tailored deceptions that specifically target their tactics and techniques.

Multilayered Deception

- **Layered Ecosystems:** Develop a multilayered deception ecosystem where fake data, systems, and interactions exist across multiple points in and outside the network. This approach creates a web of deceptions that makes it harder for attackers to identify the true target.
- **Integrated Deception:** Integrate deceptive elements into every layer of the network, from endpoints to servers, creating a comprehensive deception strategy that covers all potential attack vectors.

As deception technology evolves, leveraging AI and creating multilayered ecosystems can enhance the complexity and effectiveness of these tactics.

DECEPTION AS A SOCIETAL TOOL

- **Public Awareness Campaigns:** Use deception on a broader scale to deter cybercrime. Public awareness campaigns can create a perception of higher risk for attackers, making them think twice before launching attacks.
- **Misinformation Campaigns:** Deploy misinformation campaigns to confuse and mislead cybercriminals, making it difficult for them to plan and execute attacks effectively.

FOCUS ON SPECIFIC ATTACKER TYPES

State-Sponsored Attackers

- **Tailored Deception Tactics:** Develop sophisticated deceptions that mimic high-value targets likely to attract state-sponsored attackers. These deceptions can include fake research data, intellectual property, or governmental communications.
- **Advanced Honeypots:** Create advanced honeypots that simulate high-security environments, providing state-sponsored attackers with a seemingly valuable target while gathering intelligence on their methods.

Financially Motivated Cybercriminals

- **Financial Deceptions:** Use deceptions that mimic financial systems, such as fake bank accounts, credit card information, and payment processing systems. This lures financially motivated attackers and helps identify their tactics.
- **Ransomware Traps:** Develop ransomware traps that attract attackers with fake data they believe is valuable, allowing security teams to analyze their encryption methods and develop countermeasures.

Different types of attackers require tailored deception tactics. Understanding the specific motivations and methods of state-sponsored attackers versus financially motivated cybercriminals can help design more effective deceptions.

DECEPTION AND SOCIAL ENGINEERING

Countering Social Engineering Attacks

- **Fake Social Media Profiles:** Create fake social media profiles that appear genuine to identify and monitor attackers attempting social engineering or phishing attacks. These profiles can interact with attackers, gathering intelligence on their tactics.
- **Phishing Simulations:** Use deception to train employees to be more resistant to phishing attempts. This involves creating simulated phishing emails and monitoring how employees respond, providing targeted training based on the results.

Training and Awareness

- **Gamified Training Scenarios:** Develop gamified training scenarios where employees participate in mock phishing attacks and other social engineering tactics. Successful identification and reporting of these attempts can be rewarded, reinforcing positive security behavior.
- **Interactive Workshops:** Conduct interactive workshops where employees learn about social engineering tactics and practice identifying and responding to these threats in a controlled environment.

DECEPTION AND INSIDER THREATS

Mitigating Insider Threats

- **Planting False Information:** Plant false information within internal systems to detect malicious insider activity. This can involve fake data files, credentials, or access logs that, when accessed, trigger alerts to the security team.
- **Behavioral Monitoring:** Use deception to monitor and analyze insider behavior, identifying patterns that may indicate malicious intent. This can include setting up fake administrative access points to see if insiders attempt to gain unauthorized access.

Creating a Culture of Security Awareness

- **Regular Training:** Provide regular training on insider threat detection and mitigation, ensuring employees understand the importance of maintaining a secure environment.
- **Security Awareness Campaigns:** Launch security awareness campaigns that emphasize the risks and consequences of insider threats, fostering a culture of vigilance and responsibility.

Addressing insider threats requires a combination of deceptive tactics and a strong culture of security awareness within the organization.

THE FUTURE OF DECEPTION TECHNOLOGIES

Emerging Trends

- **Blockchain for Tamper-Proof Honeypots:** Explore using blockchain technology to create tamper-proof honeypots. This ensures that data collected within these

honeypots remains secure and immutable, providing reliable intelligence on attacker behavior.
- **AI-Driven Adaptive Deceptions:** Leverage advancements in AI to create even more sophisticated and adaptive deceptions. AI can continuously learn from attacker behavior and adjust deception strategies in real time.

Sophisticated and Adaptive Deceptions

- **Real-Time Adjustments:** AI-driven deceptions can adjust in real time based on attacker interactions, making it more challenging for attackers to identify and avoid traps.
- **Predictive Deception Strategies:** Use AI to predict future attack patterns and develop proactive deception strategies that stay ahead of evolving threats.

Challenges in Implementation

- **Keeping Pace with Evolving Threats:** Constantly updating and refining deception techniques is necessary to keep pace with the rapidly evolving threat landscape.
- **Developing Counter-Deception Strategies:** Attackers may develop counter-deception strategies, requiring continuous innovation and adaptation of deceptive tactics.

As we look to the future, emerging technologies like blockchain and AI will play a crucial role in developing more sophisticated and adaptive deception strategies, though challenges in implementation must be considered.

DECEPTION AND ACTIVE DEFENSE

Moving Beyond Detection

- **Active Defense Measures:** Deception can be used not just for passive intelligence gathering but also for active defense measures. This involves disrupting attacker tools, redirecting attacks, or feeding attackers with manipulated data that throws them off course. For instance, booby-trapped information can damage the attacker's systems or tools when accessed.
- **Disrupting Attackers:** Use deception to disrupt attacker operations by providing fake data that appears real but is designed to be difficult to process or use. This can slow down or halt their progress.

Deception-Triggered Countermeasures

- **Automated Countermeasures:** Design deceptions that trigger automated countermeasures when attackers interact with them. This can include deploying firewalls, quarantining infected systems, or launching denial-of-service attacks against the attacker's own infrastructure.
- **Dynamic Response:** Implement a dynamic response system where the deception environment can automatically adapt and escalate defensive measures based on the severity of the attack.

Moving beyond passive detection, active defense measures can use deception to disrupt attacker operations and trigger automated countermeasures.

DECEPTION AND SUPPLY CHAIN SECURITY

Deception in Third-Party Systems

- **Integrated Deception Techniques:** Incorporate deception techniques into third-party software and services to identify vulnerabilities and malicious actors within the supply chain. This can involve planting hidden code that monitors for suspicious behavior or creating fake Application Programming Interfaces (APIs) that lure attackers and expose their tactics.
- **Securing Supply Chains:** Use deception to monitor and secure supply chain interactions, ensuring that any attempt to exploit third-party relationships is detected and mitigated.

Honeynet as a Service (HaaS)

- **Outsourced Deception Deployment:** Introduce the concept of Honeynet as a Service (HaaS), where organizations can outsource the deployment and management of honeypots to specialized security providers. This allows smaller organizations to benefit from deception techniques without the need for significant in-house expertise.
- **Scalable Solutions:** Provide scalable deception solutions that can be tailored to the needs of different organizations, ensuring that even those with limited resources can implement effective deception strategies.

Deception techniques can also enhance supply chain security by identifying vulnerabilities in third-party systems and offering scalable solutions like HaaS.

DECEPTION AND REGULATORY COMPLIANCE

Meeting Compliance Requirements

- **Demonstrating Compliance:** Use deception techniques to demonstrate compliance with data privacy regulations like General Data Protection Regulation (GDPR) or California Consumer Privacy Act (CCPA). This involves using deception to test the effectiveness of data security controls and identify potential weaknesses before regulators do.
- **Proactive Testing:** Implement proactive testing measures that use deception to ensure data protection measures are effective and compliant with regulatory standards.

Deception for Regulatory Sandboxes

- **Controlled Testing Environments:** Use deception in regulatory sandboxes, which are controlled environments where businesses can test new technologies and processes without compromising real data. Deception techniques can create realistic attack scenarios within the sandbox to assess the effectiveness of security controls.
- **Safe Innovation:** Encourage innovation by providing a safe environment where new security measures can be tested and refined using deceptive tactics to simulate real-world attacks.

Deception can play a crucial role in meeting regulatory compliance and fostering safe innovation through controlled testing environments.

DECEPTION AND INTERNATIONAL COOPERATION

Global Deception Networks

- **International Intelligence Sharing:** Explore the possibility of creating international deception networks where countries share intelligence gathered through honeypots and other deception techniques. This can help track down cybercriminals operating across borders and develop more coordinated defensive strategies.
- **Collaborative Defense:** Foster international collaboration by sharing best practices and developing joint deception strategies that enhance global cybersecurity efforts.

Deception for Attribution

- **Identifying Attackers:** Use deception techniques to help with attribution, the process of identifying the source of a cyberattack. This can involve planting digital breadcrumbs within deceptive environments that lead back to the attacker's infrastructure.
- **Attribution Accuracy:** Improve the accuracy of attribution efforts by using deception to gather detailed intelligence on attacker methods and origins, providing clearer evidence for law enforcement and diplomatic responses.

International cooperation through global deception networks and improved attribution techniques can strengthen global cybersecurity efforts.

DECEPTION AND THE HUMAN ELEMENT

Social Engineering Countermeasures

- **Fake Training Materials:** Create fake training materials that attackers might target, allowing organizations to monitor and analyze attempts to exploit these materials. This can help identify attackers and their methods.
- **Disrupting Attacker Communication Channels:** Use deception to disrupt attacker communication channels. This can involve creating fake online forums or chat groups where attackers believe they are communicating with fellow cybercriminals but are actually being monitored.

Deception Fatigue

- **Understanding Deception Fatigue:** Introduce the concept of "deception fatigue," where the overuse of deception techniques makes them less effective over time as attackers become wary and more adept at identifying them.
- **Mitigating Deception Fatigue:** Explore strategies to mitigate deception fatigue, such as rotating deceptions, varying the complexity of deceptive elements, and using deceptions sparingly to maintain their effectiveness.

Addressing the human element in deception strategies, including social engineering countermeasures and managing deception fatigue, is crucial for maintaining the effectiveness of these techniques.

LOOKING TO THE FUTURE

AI-Generated Deceptions

- **Real-Time Generation:** Explore the potential of AI to generate entirely new and creative deception techniques in real time. AI can analyze current attack patterns and create deceptions that are tailored to the specific tactics and strategies of attackers.
- **Unpredictable Deceptions:** Use AI to create deceptions that are unpredictable and constantly changing, making it extremely difficult for attackers to develop counter-deception strategies.
- **AI-Driven Adaptive Deceptions:** Taking deception to the next level

Code Snippet: A Glimpse into Adaptive Honeypot Behavior

```
import random

def adapt_honeypot(attacker_behavior):
    if attacker_behavior["file_access_attempts"] > 5:
        # Rename sensitive files
        rename_files()
    if attacker_behavior["network_scan_detected"]:
        # Alter network traffic patterns
        modify_traffic()
    # ... (other adaptation rules)

# ... (main honeypot logic)
```

Algorithm Explanation: The Brains Behind Adaptive Deception

- **Reinforcement Learning:** AI learns by trial and error, receiving rewards for successful deceptions.
- **Decision Trees:** AI creates a tree-like model of decisions and their consequences to choose the best deception tactic.

Technical Challenges: The Balancing Act of Adaptation

- **Overfitting:** AI might become too specialized in deceiving a particular attacker, failing against others.
- **Continuous Learning:** Necessitates ongoing data collection to keep the AI model updated with evolving attacker tactics.

Challenges and Opportunities

- **Future Challenges:** Discuss potential future challenges in implementing deception technologies, such as the need for constant updates to keep up with evolving threats and the risk of attackers developing counter-deception strategies.
- **Innovative Solutions:** Highlight innovative solutions and cutting-edge research in the field of cybersecurity. Show how organizations can stay ahead of the curve by adopting new technologies and methodologies.

Reflection Questions

- How can your organization integrate AI-generated deceptions to stay ahead of cyber threats?

- What measures can you implement to mitigate deception fatigue within your security strategy?

Looking to the future, AI-generated deceptions and addressing future challenges will be key to maintaining effective and innovative deception strategies.

DECEPTION CASE STUDIES – GAMIFICATION

Case Study: Gamified Phishing Training

- **Scenario:** An organization implemented a gamified phishing training program where employees received simulated phishing emails regularly. Employees who identified and reported phishing attempts earned points and badges, creating a competitive and engaging environment.
- **Outcome:** The gamified approach significantly improved employee awareness and response to phishing attempts. The organization saw a 50% reduction in successful phishing attacks and a marked increase in employee engagement with security training.

Case Study: Deception in Action

- **Scenario:** A financial institution used deception to create fake wire transfer systems that mimicked real ones.
- **Outcome:** Attackers attempted to compromise the fake systems, revealing their methods and tools. The institution used this intelligence to bolster their real wire transfer systems and improve their fraud detection mechanisms.

Practical case studies illustrate the real-world application and effectiveness of deception techniques, providing valuable insights for organizations looking to implement similar strategies.

METRICS AND ROI OF DECEPTION TECHNIQUES

Measuring the effectiveness of deception strategies

Attacker Engagement Metrics

- **Number of Interactions:** Track the number of attackers interacting with honeypots and other deceptive elements.
- **Time Spent Engaged:** Monitor the time attackers spend engaged with deceptions, indicating wasted effort.
- **Types of Attacks:** Analyze the types of attacks launched against deceptions to understand attacker tactics.

Improved Security Posture Metrics

- **Reduction in Successful Attacks:** Measure the reduction in successful attacks on real systems after deploying deceptions.
- **Vulnerabilities Identified:** Track the number of new vulnerabilities identified through intelligence gathered from attacker interactions with deceptions.
- **Effectiveness of Countermeasures:** Monitor the effectiveness of deception-triggered countermeasures, such as automated threat isolation or attacker disruption tactics.

Human Behavior Metrics

- **Phishing Training Effectiveness:** Evaluate the effectiveness of gamified phishing training exercises in improving employee awareness and response rates.
- **Suspicious Activity Reporting:** Track the number of suspicious activities reported by employees after deception-based social engineering awareness campaigns.

Calculating the ROI of deception

Cost of Deception Implementation

- **Acquisition and Deployment:** Factor in the cost of acquiring and deploying deception tools (e.g., honeypot software, training).
- **Ongoing Maintenance:** Include the cost of ongoing maintenance and management of deception environments.
- **Additional Personnel:** Consider any additional personnel required to monitor and analyze data from deceptions.

Benefits of Deception

- **Cost Savings:** Quantify the potential cost savings from prevented cyberattacks (e.g., data breach costs, downtime costs).
- **Improved Threat Intelligence:** Estimate the value of improved threat intelligence gathered through deception techniques.
- **Reduced Wasted Time:** Consider the cost savings from reduced employee time wasted on responding to phishing attempts or social engineering scams.

ROI Calculation

- **Basic ROI Ratio:** Divide the quantified benefits by the cost of implementation to get a basic return on investment (ROI) ratio.

Additional Considerations

- **Varying ROI:** The ROI of deception will vary depending on the organization's specific security posture, industry, and risk tolerance.
- **Layered Security Strategy:** Deception is often most effective as part of a layered security strategy, making it difficult to isolate its individual ROI.
- **Intangible Benefits:** The intangible benefits of deception, such as improved deterrence and attacker confusion, should also be factored into the overall security value proposition.

By employing these metrics and ROI considerations, organizations can make a more informed decision about adopting deception techniques and measure their effectiveness in bolstering their overall cybersecurity posture.

CONCLUSION: RESPONSIBLE IMPLEMENTATION

Deception techniques, when implemented responsibly, can be a powerful tool in your cybersecurity arsenal. However, it's vital to ensure compliance with legal frameworks and adhere to ethical boundaries. Transparency and clear communication with law enforcement are essential to avoid any unintended consequences. By understanding the different techniques, their effectiveness, and the ethical considerations involved, you can leverage deception responsibly to gain a significant advantage in the ongoing battle against cyber threats. Remember, deception is most effective when used as part of a layered defense approach that includes traditional security measures like firewalls and IDS.

Reflection Questions

- How can your organization ensure responsible implementation of deception techniques?
- What steps can you take to balance ethical considerations with the need for effective security measures?

Key Takeaways

- **Comprehensive Defense:** Deception techniques can catch attackers regardless of their tactics, techniques, and procedures (TTPs) or zero-day exploits.
- **Proactive Measures:** These strategies can slow down attackers, create deterrence, capture true positives, collect forensics on active attackers, and develop engaging stories and campaigns around your organization's crown jewels.

CYBER CHUCKLES

- **Why did the red teamer start a magic show?** They wanted to master the art of deception!
- **Why don't cyber defenders trust their computers?** Too many "traps" and "tricks"!
- **Why did the attacker bring a compass?** To navigate through the maze of deception!
- **How do hackers stay in shape?** By running into fake networks and lifting false data!
- **Why did the AI-powered honeypot join the chess club?** To learn more about "strategic moves" against attackers!
- **Why was the phishing email disappointed?** It fell for a decoy email account!

By exploring these new perspectives, cyber deception can become a more versatile and powerful tool for cybersecurity professionals.

Chapter 11

Red teaming, blue teaming, purple teaming, and yellow teaming

A holistic approach to security assessments

INTRODUCTION

- **Escalating Cyber Threat Landscape:** The complexity and frequency of cyber threats are growing, making proactive security measures essential for protecting organizational assets.
- **Role of Security Testing and Exercises:** Security testing and exercises are crucial for uncovering vulnerabilities and improving overall security posture.
- **Introducing the Teaming Approaches:** This chapter explores red, blue, purple, and yellow teaming as complementary strategies for comprehensive security assessment.

SECURITY TESTING APPROACHES

Red teaming

- **Beyond Penetration Testing:** Red teaming goes beyond traditional penetration testing by simulating a broader range of attacker behaviors, including social engineering, supply chain threats, and insider threats.
- **Red Teaming as a Service (RTaaS):** RTaaS is becoming popular for organizations that lack in-house red team expertise, offering comprehensive assessments through external experts.
- **Ethical Considerations:** Establish clear rules of engagement, ethical guidelines, and legal compliance to ensure that exercises are conducted responsibly.
- **Historical Context:** Red teaming has its roots in military exercises and has evolved to address modern cybersecurity challenges.
- **Red Teaming as Threat Hunting:** Red teaming tactics can be adapted for proactive threat hunting, continuously searching for vulnerabilities and potential attacker behavior.
- **Red Teaming with Artificial Intelligence (AI):** AI can automate tasks, personalize attack simulations, and identify previously unknown vulnerabilities, making red teaming more efficient.
- **Metrics and Measurement:** Examples of metrics include the number of vulnerabilities identified, exploitation success rates, and time to compromise.
- Case Studies
  1. **Global Financial Institution:** A global financial institution used red teaming to uncover a critical vulnerability in their online banking system. The exercise involved simulating a sophisticated phishing attack that bypassed initial defenses and highlighted weak spots in user training and incident response protocols. As a

DOI: 10.1201/9781003562184-16

result, the institution improved its security awareness programs and strengthened its incident response plans.
2. **Healthcare Provider:** A healthcare provider discovered weaknesses in their supply chain security through a red teaming exercise. The red team simulated an attack on a third-party vendor, revealing that the vendor's access controls were insufficient. This led to a comprehensive review and upgrade of the provider's vendor management practices, including stricter access controls and regular security audits.

Blue teaming

- **Automation in Blue Teaming:** Automation tools enhance blue team efficiency, particularly in threat detection and incident response. Tools like Security Information and Event Management (SIEM) systems help automate the collection, analysis, and correlation of security event data.
- **Blue Teaming Games and Simulations:** Gamified simulations train blue teams to handle real-world attacks in controlled environments, improving their preparedness. These exercises can range from tabletop simulations to full-scale mock incidents.
- **Metrics and Measurement:** Metrics such as Mean Time to Detect (MTTD) and Mean Time to Respond (MTTR) are essential for evaluating blue team performance and tracking progress.
- **Future Trends:** AI and machine learning (ML) are emerging trends that accelerate threat detection and response, making blue teams more effective. Predictive analytics can forecast potential security incidents based on historical data.
- **Blue Teaming with Machine Learning (ML):** ML can analyze vast amounts of security data quickly, improving incident detection and response times. For example, anomaly detection algorithms can identify unusual patterns of behavior that may indicate a security breach.
- **Blue Teaming and Deception Technologies:** Deception technologies simulate realistic attack scenarios and detect attacker movements, enhancing blue team exercises. Techniques such as honeypots, honeytokens, and deception grids can mislead attackers and gather intelligence on their methods.
- **Metrics and Measurement:** Examples include incident response time, accuracy of threat detection, and number of false positives/negatives.
- **Case Studies**
  1. **Tech Company:** A tech company improved their MTTR by 40% through the implementation of ML algorithms in their blue team operations. By integrating ML into their SIEM system, the blue team could prioritize alerts more effectively and respond to high-risk incidents faster.
  2. **E-Commerce Platform:** An e-commerce platform used deception technologies to successfully detect and mitigate a sophisticated phishing attack. The deployment of decoy email accounts and fake credentials helped identify the attackers' methods and prevented data breaches.

The collaborative approach: purple teaming

- **Definition and Benefits:** Purple teaming integrates red and blue team activities, fostering collaboration and knowledge sharing for enhanced security. This approach helps bridge the gap between offensive and defensive strategies.
- **Purple Teaming for OT Security:** Adapt purple teaming for operational technology (OT) environments to address security concerns related to physical infrastructure.

This includes testing the resilience of industrial control systems (ICS) and Supervisory Control and Data Acquisition (SCADA) systems against cyber threats.

- **Continuous Purple Teaming:** Continuous collaboration between red and blue teams throughout the security lifecycle enhances overall security posture. This involves regular, iterative engagements rather than one-off exercises.
- **Challenges of Implementation:** Overcoming silos and fostering a culture of collaboration can be challenging but is essential for effective purple teaming. Leadership support and clear communication channels are critical for success.
- **Purple Teaming for Cloud Security:** Purple teaming is crucial for securing cloud environments, where shared responsibility models require close collaboration between security teams. This includes testing cloud configurations, access controls, and incident response procedures.
- **Metrics and Measurement:** Track collaboration effectiveness, number of vulnerabilities remediated, and improvements in detection and response times.
- **Case Studies**
  1. **Manufacturing Firm:** A manufacturing firm significantly reduced downtime and improved security resilience through continuous purple teaming exercises. The firm conducted regular red-blue team engagements, leading to the discovery and remediation of several critical vulnerabilities in their production network.
  2. **Retail Company:** A retail company strengthened its cloud security posture by integrating purple teaming into its regular security operations. The collaborative efforts helped identify misconfigurations and optimize their cloud security controls, reducing the risk of data breaches.

Introducing yellow teaming

- **Definition and Purpose:** Yellow teaming analyzes the potential negative impacts of security measures on business operations, ensuring they align with organizational goals. This approach focuses on the balance between security and business efficiency.
- **Cost-Benefit Analysis with Real-World Examples:** Yellow teaming helps organizations avoid costly and disruptive security measures through thorough cost-benefit analyses. For instance, evaluating the operational impact of implementing multifactor authentication (MFA) across all business units.
- **Yellow Teaming and Security ROI:** Maximizes security investment return on investment (ROI) by ensuring security measures contribute to overall business objectives. This includes prioritizing security initiatives that provide the most significant benefits with the least disruption.
- **Integrating with Security Risk Management:** Yellow teaming can be integrated with existing security risk management frameworks for a holistic approach. This ensures that security decisions consider both technical and business risks.
- **Yellow Teaming and Security Culture:** Fosters a strong security culture by raising awareness of potential negative impacts from security measures. Engaging employees in the decision-making process helps build a culture of security ownership and accountability.
- **Quantifying Yellow Teaming Impact:** Includes methods for quantifying the positive impact of yellow teaming, such as measuring reductions in security incidents or operational disruptions. Metrics can include reduced downtime, increased user satisfaction, and compliance with regulatory requirements.

- **Yellow Teaming in Non-Technical Contexts:** Apply yellow teaming principles to analyze the impacts of new policies or procedures in various organizational areas. For example, assessing the impact of a new remote work policy on security and productivity.
- **Yellow Teaming and Regulatory Compliance:** Ensures security measures comply with evolving regulations, minimizing legal risks and penalties. This includes reviewing and updating security policies to meet regulatory standards.
- **Yellow Teaming for Mergers and Acquisitions (M&A):** Assesses the potential security and operational impacts of integrating different IT systems during mergers and acquisitions. This involves evaluating the security posture of the acquired company and planning for a seamless integration.
- **Metrics and Measurement:** Examples include reduced operational disruptions, improved business continuity, and compliance with regulatory requirements.
- **Case Studies**
  1. **Financial Services Company:** A financial services company used yellow teaming to avoid a costly and disruptive security implementation. The yellow team assessed the impact of a proposed data encryption solution and recommended a phased implementation that minimized disruption and maintained compliance.
  2. **Multinational Corporation:** A multinational corporation applied yellow teaming during a merger, ensuring seamless integration of IT systems and maintaining business continuity. The yellow team identified potential integration risks and developed mitigation strategies, leading to a smooth transition.

KEY RESPONSIBILITIES

Risk and Harm Analysis

- **Steps for Identifying Potential Negative Impacts:** Detail the process for identifying operational disruptions, financial costs, and compliance issues.
- **Conducting Cost-Benefit Analyses:** Provide a framework for thorough cost-benefit analyses to ensure security measures are justified. This involves evaluating the potential benefits of a security measure against its impact on business operations.

Impact Assessment

- **Evaluating Operational Impact:** Assess impacts on daily workflows and user experience to ensure security measures do not hinder business operations. This includes conducting surveys and focus groups to gather feedback from employees.
- **Business Continuity Implications:** Consider the importance of maintaining business continuity when implementing security measures. This involves planning for potential disruptions and developing contingency plans.
- **Legal and Compliance Considerations:** Evaluate legal and compliance implications to ensure security measures adhere to regulations. This includes consulting with legal and compliance teams to review security policies and procedures.

Stakeholder Engagement

- **Consulting Various Departments:** Involve business units, legal, HR, and IT to gather diverse perspectives and ensure comprehensive assessments. This ensures that all relevant stakeholders are considered in the decision-making process.

- **Establishing Feedback Mechanisms:** Collect and incorporate stakeholder feedback to refine security measures and address concerns. This includes regular meetings, surveys, and feedback sessions to gather input from employees.

Implementation plan

Establish the Yellow Team

- **Necessary Expertise:** Describe the required skills in risk management, business operations, legal, and security to form an effective yellow team. This includes identifying key stakeholders and subject matter experts from various departments.

Develop Analysis Framework

- **Impact Assessment Matrix and Risk Scoring System:** Explain how to create these tools for evaluating the impact of security measures. This involves developing criteria for assessing the impact on business operations and assigning risk scores based on the likelihood and severity of potential impacts.

Conduct Impact Assessments

- **Methods Used:** Detail scenario analysis and stakeholder workshops to thoroughly assess potential impacts. This includes conducting tabletop exercises and simulations to evaluate the impact of security measures on business operations.

Review and Recommendation Process

- **Harm Mitigation Strategies:** Develop strategies to mitigate identified risks and provide balanced recommendations. This involves collaborating with stakeholders to develop practical solutions that minimize disruption and maintain business continuity.

Continuous Monitoring and Evaluation

- **Post-Implementation Reviews:** Conduct ongoing reviews and establish feedback loops for continuous improvement. This includes regular assessments and updates to security measures based on feedback and changing business needs.

Example activities

Policy Impact Analysis

- Provide a concrete scenario where a policy change is evaluated for its impact on operations and security. For example, assessing the impact of implementing a new data privacy policy on business processes and customer interactions.

Technology Implementation Review

- Discuss a real-world example of assessing the impact of implementing new security technology, highlighting practical considerations. For instance, evaluating the impact of deploying a new endpoint security solution on employee productivity and network performance.

Benefits of yellow teaming

Minimized Negative Impacts

- **Examples:** Provide specific examples of how yellow teaming has minimized operational disruptions, ensuring business continuity. For example, avoiding the implementation of a disruptive security measure that could have caused significant downtime.

Improved Business Continuity

- **Case Studies:** Highlight instances where yellow teaming contributed to better business continuity, demonstrating its effectiveness. For example, ensuring a seamless integration of IT systems during a merger, minimizing disruption to business operations.

Enhanced Stakeholder Buy-In

- **Real-World Examples:** Show how involving stakeholders early in the process has led to greater acceptance and successful implementation of security measures. For example, engaging employees in the decision-making process to ensure buy-in and compliance with new security policies.

CONCLUSION

- **Key Takeaways:** Summarize the importance of understanding and leveraging red, blue, purple, and yellow teaming for a comprehensive and effective security strategy. Emphasize the benefits of a holistic approach to security assessments and the importance of collaboration and continuous improvement.

NOVEL PERSPECTIVES

Improvements

1. **Integration of Emerging Technologies**
    - **AI and Machine Learning:** Explore how AI and ML can enhance red, blue, purple, and yellow teaming efforts. Discuss how these technologies can be used to simulate more sophisticated attacks, improve detection and response, and analyze the impact of security measures on business operations.
2. **Real-World Scenarios**
    - **Detailed Case Studies:** Include more in-depth case studies for each teaming approach, showing real-world applications and outcomes. This can help illustrate the practical benefits and challenges of each strategy.
    - **Incident Narratives:** Present narratives of actual security incidents where each teaming approach played a crucial role in mitigating the threat or recovering from an attack.
3. **Interdisciplinary Collaboration**
    - **Cross-Functional Teams:** Emphasize the importance of creating cross-functional teams that include not only IT and security professionals but also members from operations, legal, HR, and other relevant departments.
    - **Cultural Integration:** Discuss strategies for fostering a culture of collaboration and continuous improvement across different teams.

4. **Focus on Continuous Improvement**
    - **Feedback Loops:** Highlight the importance of continuous feedback and improvement in security practices. Show how regular reviews and updates based on lessons learned can enhance the effectiveness of each teaming approach.
    - **Adaptive Strategies:** Discuss the need for adaptive security strategies that can evolve with changing threats and business environments.
5. **Global and Ethical Considerations**
    - **Global Collaboration:** Explore the role of international collaboration in improving security assessments. Discuss how organizations can work together across borders to share knowledge and resources.
    - **Ethical Implications:** Delve deeper into the ethical considerations of each teaming approach, especially when it comes to red teaming and the use of social engineering tactics.
6. **Visualization and Engagement**
    - **Infographics and Diagrams:** Use visual aids to illustrate the processes and interactions between red, blue, purple, and yellow teams. This can make complex concepts more accessible and engaging.
    - **Interactive Elements:** If possible, suggest interactive elements like online simulations or gamified training exercises that readers can try to better understand the concepts.

A SWITCH OF PERSPECTIVE

1. Hybrid Teaming Models
    - **Integrating Multiple Approaches:** Propose a hybrid teaming model that combines elements of red, blue, purple, and yellow teaming into a cohesive framework. Discuss how this integrated approach can provide a more comprehensive security assessment.
    - **Customizable Frameworks:** Offer customizable frameworks that organizations can tailor to their specific needs and threat landscapes.
2. **Psychological and Behavioral Insights**
    - **Human Factors:** Incorporate insights from psychology and behavioral science to understand how human behavior impacts security. Discuss how understanding cognitive biases and decision-making processes can improve the effectiveness of security teams.
    - **User-Centric Security:** Emphasize the importance of user-centric security measures that take into account the needs and behaviors of end users. Show how yellow teaming can help design security solutions that are user-friendly and effective.
3. **Future-Forward Thinking**
    - **Anticipating Future Threats:** Encourage readers to think about future threats and how current teaming approaches can be adapted to address them. Discuss emerging technologies like quantum computing and their potential impact on security assessments.
    - **Innovative Solutions:** Highlight innovative solutions and cutting-edge research in the field of cybersecurity. Show how organizations can stay ahead of the curve by adopting new technologies and methodologies.

CYBER CHUCKLES

- **Why did the red teamer bring a fishing rod to work?** They heard it was a great way to "phish" for credentials!

- **What's a blue teamer's favorite type of music?** Heavy "metrics" – because they're always tracking those numbers!
- **Why did the purple team go on a team-building retreat?** To break down those silos and build some real "silo-darity!"
- **How does a yellow teamer make a tough decision?** With a strong "cost-benefit" brew – analyzing every last drop!
- **What did the AI say to the hacker?** "I'm watching you with my machine learning eyes!"
- **Why did the cyber analyst start gardening?** They wanted to learn how to "decompose" those logs more efficiently!

Chapter 12

Is Cyberfantastic still fantastic?

Re-evaluating the future of cybersecurity

Spoiler Alert!

- Buckle up for a journey through the ever-evolving world of cybersecurity.
- This chapter explores "Cyberfantastic," a future where security is proactive and, well, fantastic!
- But wait, there's a catch! Cyberfantastic comes with its own set of challenges.
- Learn how to be a cybersecurity pioneer and navigate the road to a "cyberfantastic" future (complete with a roadmap, because who wants to get lost?).

PREFACE

In the dynamic domain of cybersecurity, the conventional pursuit has been to achieve unassailable protection, leading us to construct near-inviolable digital barriers and predict every possible threat. However, emerging insights suggest a transformative perspective that diverges from this established belief.

Consider a prominent high-frequency trading entity, constantly immersed in the volatility of financial markets. Instead of merely reinforcing its defenses, this entity introduces controlled disruptions to its trading system – intentional delays, calculated glitches, and even simulated cyberattacks. This proactive approach, while appearing counterintuitive, is grounded in a foundational cybersecurity insight.

Channeling disruption to craft strength

This methodology extends beyond standard system resilience. Drawing inspiration from the tenets of Judo, it perceives disruptions not as mere threats, but as opportunities for advancement. Instead of purely countering adversarial forces, the system leverages them, propelling its own evolution and fortification.

The shift from predictability to adaptability

The traditional cybersecurity framework prioritizes prediction, emphasizing preemptive measures against perceived threats. However, in an increasingly complex digital landscape, this reactive mindset may not be sufficient. The trading institution, in our example, emphasizes adaptability. By consciously evolving and refining its mechanisms, it not only withstands disruptions but leverages them for augmentation.

Extend this philosophy to a globally operating healthcare institution, which, despite its robust cybersecurity measures, introduces controlled vulnerabilities to its electronic health record (EHR) system.

DOI: 10.1201/9781003562184-17

Employing vulnerabilities as catalysts

Emulating the aforementioned trading institution and drawing from Judo's tenets, this healthcare entity transforms perceived vulnerabilities into growth catalysts. The introduction of controlled vulnerabilities serves not as a risk but as a tool for continuous improvement.

Adaptability as a core strategy

In cybersecurity, while predictability has its merits, adaptability offers a proactive approach. The healthcare institution prioritizes dynamic, responsive strategies over static, predictive ones, resulting in an EHR system that's not just secure but constantly evolving.

This innovative paradigm is not restricted to a single sector but has implications across industries. Whether it's an energy firm introducing vulnerabilities to its smart grid or an educational institution optimizing its student information systems, the principle remains consistent.

As we traverse the intricate pathways of the digital era, it's pivotal to acknowledge that groundbreaking solutions often originate from unconventional ideas. In this new frontier, controlled vulnerabilities could be pivotal in reshaping cybersecurity. The journey into this novel territory has commenced, and the ensuing revelations promise to be transformative.

MOTIVATION: EMBRACING NEW APPROACHES FOR A SECURE DIGITAL FUTURE

In the evolving landscape of cybersecurity, I've always been intrigued by the idea of systems that don't just survive attacks but grow stronger because of them. This vision, where our digital defenses not only withstand threats but flourish in their midst, represents a transformative shift in our field.

We must face a clear truth: the realm of cybersecurity is ever-changing. New vulnerabilities surface continually, and the axiom "it's not a matter of if but when" aptly captures the industry's sentiments. This has often set us in a seemingly endless chase, where adversaries devise new exploits and we, in turn, seek to patch and strengthen against them.

However, let's consider a more proactive model. What if, instead of always playing catch-up, our systems learned and grew stronger with every challenge they faced? This vision encapsulates the essence of Cyberfantastic.

Imagine a digital ecosystem that doesn't just respond to threats but leverages them to bolster its defenses. Picture a cybersecurity framework that thrives amidst adversities, viewing each threat as an opportunity for growth and refinement. This is the heart of Cyberfantastic, a philosophy with the potential to revolutionize our conventional notions of risk management.

For far too long, we've leaned heavily on traditional risk management tactics, often finding ourselves lagging behind the fast-paced evolution of threats. Our security models, largely reactive in nature, primarily focus on mitigating risks after they've been identified. And while these measures are crucial, they often leave us a step behind our adversaries.

Cyberfantastic offers a way forward. It represents more than just a new strategy; it signifies a paradigm shift in cybersecurity. It challenges us to shift our perspective, moving from pure defense to growth and adaptability. It inspires us to welcome challenges, learn from them, and come out stronger.

This concept, Cyberfantastic, might just be the answer to some of the long-standing risk management challenges plaguing our industry. Rather than continually reacting to new threats, it paves the way for building systems that proactively evolve in response. Instead of being consistently on the defensive, it allows us to turn the tables, using adversarial tactics to fuel our own growth.

INTRODUCTION

The journey of cybersecurity evolution charts a transformative path, ushering organizations from states of vulnerability to unparalleled resilience and, ultimately, to a Cyberfantastic posture. This progressive trek traverses' specific milestones, each hallmarked by increasing degrees of strategic foresight, risk management, and technological prowess. As the digital universe intensifies in its interconnectedness and threats become more intricate, the pursuit of cybersecurity evolution serves as a luminous touchstone, directing organizations to defend their digital treasures and ensure uninterrupted operations.

Embarking on this evolution is akin to climbing a ladder of strategic readiness. Each step symbolizes a pivotal stride toward achieving advanced security states. The journey initiates at the foundational "Ad hoc security" phase, where responses are impromptu, lacking a broader strategy, resulting in patchy security efforts and occasional vulnerabilities against rising threats. As we transition to the "Compliance-based" stage, there's a shift in focus toward fulfilling regulatory mandates and reinforcing basic security protocols. Yet, this level doesn't wholly encompass the fluid nature of today's cyber threats, prompting a move to the "Risk-based" approach. Here, risks undergo meticulous evaluation and ranking, leading to the creation of bespoke security measures in line with organizational blueprints.

Understanding the nuances, hurdles, and significance of each evolutionary stage enlightens the path from vulnerability to Cyberfantastic. By wholeheartedly adopting this comprehensive trajectory, organizations set forth on a transformative voyage, solidifying their digital infrastructures, enabling them to flourish even in the face of disruptions, and culminating in an environment where their cyber systems' innate strength and flexibility overshadow traditional risk management necessities. In a digital era teeming with both challenges and opportunities, the drive toward cybersecurity evolution emerges as a torchbearer, pointing organizations to a horizon where their digital stance isn't just fortified but is genuinely Cyberfantastic.

THE EVOLUTION

In the unpredictable maze of the digital world, adapting to the ever-shifting threat terrain is no longer just an option; it's a necessity. Cybersecurity evolution is not a straightforward journey but an intricate pilgrimage through stages of vulnerability, bolstering, adaptation, resilience, and reaching the pinnacle of Cyberfantastic. This chapter dissects this transformative sojourn, shedding light on the meticulous steps organizations need to master to chart these volatile waters. From the rudimentary ad hoc security protocols to a current highpoint of Cyberfantastic, delving into how this evolutionary pathway molds the strategies, tools, and mental frameworks that enable to fortify the digital tomorrow.

Evolution Step 1: Ad hoc Security

- **Explanation:** At this nascent phase, organizations have rudimentary security infrastructure, marked by spontaneity rather than strategy.

- **Characteristics:** Elementary firewall configurations, occasional antivirus checks, and limited cognizance of looming threats.
- **Challenges:** A disjointed approach, erratic safeguards, and a heightened vulnerability to nascent threats.
- **Risk Management:** Virtually non-existent. The impromptu nature of responses amplifies risks and consequential damages.

Evolution Step 2: Compliance-Based

- **Explanation:** At this juncture, organizations pivot their focus to meeting regulatory benchmarks and industry norms, aiming to uplift their security stature.
- **Characteristics:** Laying down essential security controls, scheduling vulnerability evaluations, and framing security protocols.
- **Challenges:** A myopic compliance-centric approach might overlook threats that transcend regulatory boundaries.
- **Risk Management:** Though not pinpointed risk management, it offers a sturdier foundation for security endeavors.

Evolution Step 3: Risk-Based

- **Explanation:** Organizations begin to embrace risk analytics, spotlighting assets, chinks in the armor, and plausible repercussions.
- **Characteristics:** Undertaking risk audits, prioritizing security initiatives, activating security controls anchored to risk metrics, and crafting contingency blueprints.
- **Challenges:** Demands a profound comprehension of organizational assets juxtaposed with potential threats.
- **Risk Management:** Precise risk management becomes achievable as organizations allocate resources rooted in discerned risks.

Evolution Step 4: Proactive Security

- **Explanation:** Organizations stretch their security canopy, incorporating external stakeholders like customers, allies, third-party entities, and overseers.
- **Characteristics:** Laying down encrypted communication pathways, pooling threat intelligence, and subjecting external associates to security reviews.
- **Challenges:** Orchestrating synchronized security protocols across an array of stakeholders and navigating collective risks.
- **Risk Management:** It pivots to dual risk management, addressing both intrinsic organizational risks and those spanning external stakeholders.

Evolution Step 5: Cyberfantastic

- **Explanation:** Organizations touch a highpoint where their digital frameworks not just fend off threats but morph and fortify in reaction to them.
- **Characteristics:** Architectures that are self-regenerative, pliable, and pioneering, drawing insights from adversities and setbacks.
- **Challenges:** Crafting and sustaining such resilient ecosystems demand avant-garde tech, perpetual knowledge acquisition, and agile gameplans.
- **Risk Management:** Risk considerations diminish substantially as the system's prowess to flourish amidst adversities elevates.

Transitioning from cyber resilience to enhanced cybersecurity paradigms

Transitioning from the established realms of cyber resilience through to the cutting-edge paradigm of enhanced cybersecurity paradigms marks a significant evolution in the cybersecurity landscape. While the focus shifts to Stage 5, known as Cyberfantastic, it is crucial to recognize the importance of the preceding stages (Stages 1–4) and their role in laying the foundation for this journey (Figure 12.1).

Each of these initial stages has contributed significantly to the understanding of cybersecurity. Stage 1's ad hoc security, Stage 2's compliance-based approach, Stage 3's risk-based strategies, and Stage 4's proactive security measures have all played vital roles in fortifying digital defenses.

In the transition to Stage 5, Cyberfantastic, a paradigm shift occurs. Cyberfantastic views Cyber Antifragility not as an endpoint but as an inspiring source. It identifies that the mental shift introduced by such models like the DIE model from Sounil Yu – Distributed, Immutable, and Ephemeral – serves as a guiding compass for this journey.

The DIE model is a cornerstone of the approach in the Cyberfantastic stage. It comprises three fundamental principles:

1. **Distributed:** Embracing the concept of distribution minimizes the risk of a single point of failure. By dispersing data and resources across multiple nodes and locations, resilience is enhanced, creating a more robust defense mechanism.
2. **Immutable:** Immutability ensures the integrity of data. Once information is recorded, it becomes resistant to unauthorized alterations without creating a new version. This principle safeguards against tampering and ensures data remains uncorrupted.
3. **Ephemeral:** The strategy adopts an ephemeral nature, where data and processes exist only for the required duration. This practice reduces the window of opportunity for cyberattacks, as outdated data is not retained, thus diminishing potential vulnerabilities.

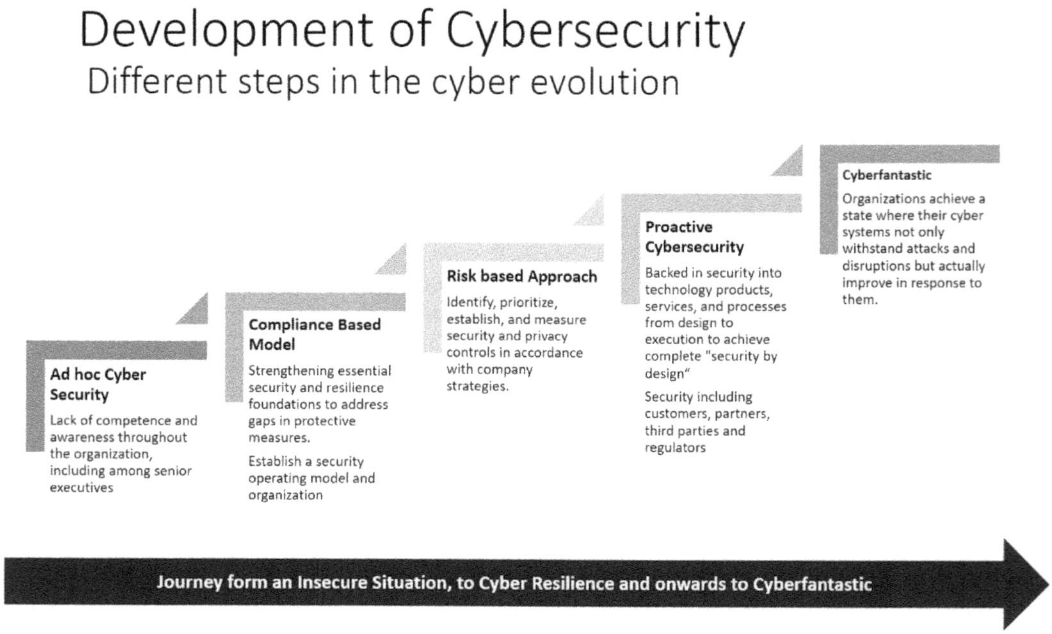

Figure 12.1 Five steps (Ad hoc, Compliance, Risk Based, Proactive, and Cyberfantastic) of cyber evolution for a company on the way to Cyberfantastic.

Importantly, Cyberfantastic acknowledges that its principles can be approached independently, even without encompassing other models or stages. These building blocks provide tangible steps toward a more robust and adaptive cybersecurity framework.

The journey from Cyber Resilience to Cyberfantastic is a testament to the evolution of cybersecurity. While each stage has its significance, Cyberfantastic, inspired by Anitfragility and the DIE model, offers a new perspective and actionable building blocks for enhancing the digital security landscape.

Distinguishing among cyber resilience, cyber antifragility, and Cyberfantastic

At the core of the digital defense strategy lies the concept of cyber resilience. It represents an organization's determination to maintain critical operations even when facing unforeseen cyberattacks. Picture it as a formidable fortress that continues to function and safeguard its core, even when under siege.

However, beyond this fortress mentality, there's "Cyber Antifragility," a concept that often takes center stage in tech discussions nowadays. In this phase, systems go beyond mere resistance to threats; they evolve and fortify themselves with each encounter with adversity.

And then, there's Cyberfantastic, a paradigm that refines and expands upon these notions. While cyber resilience focuses on defending against the storm, Cyberfantastic encapsulates the idea of harnessing the storm's fury for continuous transformation. It signifies a shift from merely withstanding challenges to leveraging them as opportunities for comprehensive enhancement.

In summary, cyber resilience equips organizations to stand strong against cyberstorms, while Cyberfantastic reshapes the defense matrix, ensuring not only the navigation of challenges but also evolution through them, ultimately turning potential vulnerabilities into sources of strength.

CYBERFANTASTIC: THE NEXT FRONTIER IN CYBERSECURITY

Cyberfantastic represents a transformative evolution in cybersecurity. Unlike traditional systems that focus on merely withstanding threats, Cyberfantastic embraces these adversities, converting them into opportunities for growth and improvement. Importantly, the foundational elements for Cyberfantastic already exist, suggesting that its realization is more a question of integration and refinement than one of invention.

What is Cyberfantastic?

At its core, Cyberfantastic is a digital ecosystem that thrives amidst disruptions. It encapsulates the essence of true cyber antifragility, marking a transition from static, reactive security measures to dynamic, proactive strategies. Systems under this paradigm don't just endure challenges; they harness them to amplify their capabilities.

Technical strategies for achieving Cyberfantastic

To transform into a Cyberfantastic entity, it's crucial to implement advanced technical strategies. These are not just bolted-on solutions; they are deeply integrated mechanisms allowing systems to enhance and adapt in response to threats:

- **Adaptive Machine Learning and AI:** Implement algorithms that can continuously learn from real-time patterns and anomalies, thus enabling autonomous adaptation.

- **Decentralized Identity Management:** Improve authentication processes, ensuring single-point credential breaches don't jeopardize the entire system.
- **Dynamic Micro-segmentation:** Proactively isolate network segments upon anomaly detection to hinder unauthorized lateral movements.
- **Autonomous Incident Response:** Utilize automated systems that refine response strategies over time, learning from each breach.
- **Predictive Analytics:** Foresee potential zero-day vulnerabilities by analyzing current digital behaviors.
- **Quantum-Resistant Techniques:** Prepare for the future with quantum-resistant cryptographic algorithms and quantum key distribution for foolproof encryption.
- **Hybrid Cloud Resilience:** Design systems that switch seamlessly between on-site and cloud operations based on threats or triggers.
- **Human-Machine Collaboration:** Adopt AI-driven tools that enhance security awareness by monitoring and flagging unusual user activities.

Amplifying privacy with onion routing

With the exponential growth of interconnected digital realms, traditional security models often lag behind. Enter onion routing – a method that exponentially boosts privacy while solidifying zero-trust architectures. By layering data encryption and routing it through various nodes, onion routing makes lateral movements within a network near-impossible for intruders.

Embracing randomness: A new paradigm in cybersecurity

Randomness, when strategically introduced into cybersecurity, disrupts predictability. This unpredictability makes it exponentially harder for attackers to decipher system behavior or exploit patterns:

- **Randomized Data Packet Routing:** Unpredictable data routes deter intruders.
- **Dynamic Port Number Allocation:** Random port allocations hinder port scanning attacks.
- **Randomized Encryption Key Generation:** Bolster defenses against brute-force attacks with random key generation.
- **Randomized Traffic Analysis:** Incorporate random traffic patterns to obscure genuine data flow.
- **Etc.**

Feedback-driven cybersecurity: The continuous learning loop

In the realm of Cyberfantastic, continuous improvement is paramount. The integration of feedback loops ensures that systems don't just recover from threats but learn and grow from them:

- **Knowledge Generation:** Accumulate and reference insights from anomalies and incidents.
- **Adaptive Adjustments:** Use accumulated knowledge to adapt and refine security protocols.
- **Human Feedback Integration:** Marry the precision of machines with the nuance and intuition of human expertise.
- **Iterative Refinement:** Continuously refine strategies based on evolving threat landscapes.

As the digital world advances, so must the approach to cybersecurity. Cyberfantastic offers a vision of the future where systems don't just resist threats but utilize them as catalysts for evolution. Organizations aiming for this future must embrace innovation, integrating both current and emerging strategies to navigate the ever-changing digital landscape confidently.

CRITICISM

In the realm of cybersecurity, the concept of Cyberfantastic could gain considerable attention for its promise of not just surviving but thriving in the face of adversity. However, as with any emerging paradigm, it is essential to subject it to critical examination. In this critique, we delve into the facets of Cyberfantastic, highlighting both its strengths and limitations, along with considerations that must be factored in when adopting this approach.

While Cyberfantastic represents a significant leap forward in bolstering digital security, it is crucial to understand that it might not serve as a universal solution. It should neither be viewed as a guaranteed remedy nor as the ultimate destination in the evolution of cybersecurity. Instead, it is a valuable component within a broader cybersecurity framework, with distinct nuances and areas of impact.

In the following discussion, some key points of consideration are explored:

- **Not a Silver Bullet:** Cyberfantastic should not be misconstrued as a silver bullet solution that guarantees absolute cybersecurity. While it is a powerful approach to enhancing resilience, it does not eliminate all cyber risks. Organizations must still implement a holistic cybersecurity strategy that includes preventive, detective, and responsive measures.
- **Not the Final Step:** Cyberfantastic represents a significant evolution in cybersecurity, but it should not be perceived as the ultimate destination. The digital landscape is continually evolving, and future threats and challenges will emerge. Organizations must continue to envision and adapt to next steps in their cyber evolution, possibly through paradigms like **Cyber Reciprocity, Cyber Symbiosis, or Cyber Resonance.**
- **Protection for IT Environment, Not Individual Systems:** Cyberfantastic primarily focuses on protecting the overall IT environment and ecosystem. While it can significantly enhance the resilience of the entire system, it doesn't address vulnerabilities at the component level, such as specific code snippets or systems. These elements may remain susceptible and require additional security measures.
- **Vulnerability of Cells (e.g., Code, Applications, Systems):** Cyberfantastic operates at a macrolevel, addressing the system as a whole. However, individual cells within the system, like for example code, applications, and systems, may still contain vulnerabilities that can be exploited. Organizations must complement Cyberfantastic with stringent security practices at the microlevel.
- **Adaptation for OT Environment:** The philosophy of Cyberfantastic, while applicable to Information Technology (IT) landscapes, needs a different approach for Operational Technology (OT) environments. OT systems, often utilized in critical infrastructure like energy grids, manufacturing plants, and transportation networks, come with distinct requirements and challenges. For instance, the real-time demands of OT systems, long equipment lifecycles, or the integration of legacy systems may not align seamlessly with Cyberfantastic principles. To implement Cyberfantastic in an OT environment, a tailored strategy considering the unique operational requirements and potential risks is indispensable.

- **Pareto Principle (80/20 Rule):** Cyberfantastic, like many other aspects of cybersecurity, might align with the Pareto Principle. This means that a significant portion of the benefits may stem from a relatively small percentage of efforts or measures. Organizations should prioritize their Cyberfantastic efforts based on risk and impact, focusing on areas where they can achieve the most significant gains.

While Cyberfantastic offers a valuable methodology to cybersecurity, it isn't a cure-all. It should be perceived as one tool among many in a comprehensive cybersecurity strategy, recognizing its limitations and the necessity for continuous adaptation to evolving threats. Moreover, its application should be context-sensitive, especially when differentiating between IT and OT environments, and prioritized based on risk and impact.

RISK MANAGEMENT IN THE CYBERFANTASTIC ERA

In the intricate domain of cybersecurity, the evolution toward Cyberfantastic presents a potential revolution, moving away from the long-standing tactic of reinforcing barriers against the myriad of threats and vulnerabilities. As organizations navigate this apex of cyber readiness, a fascinating outcome materializes: the possible diminution, though not complete eradication, of traditional risk management practices. Here are the transformative benefits underpinning this metamorphosis, illustrating how the adoption of Cyberfantastic intrinsically reduces reliance on standard risk management strategies.

1. **Favoring Adaptability Over Mere Mitigation:** Traditional risk management predominantly concentrates on detecting and mitigating potential risks via precautionary steps. In the Cyberfantastic paradigm, the emphasis evolves from mere risk aversion to developing systems that are ever adaptable. Such systems, designed to not just withstand but thrive amidst adversity, organically reduce the requirement for perpetual risk evaluations and mitigation, making room for more fluid adaptive responses.
2. **Harnessing Lessons from Challenges:** Central to Cyberfantastic is the principle of transformation through adversity. Systems, when confronted with stressors or disruptions, rebound stronger, integrating insights from the experience. This continuous adaptive learning diminishes the demand for exhaustive risk audits as systems develop a reflexive ability to pinpoint and handle emerging vulnerabilities from historical data.
3. **Expedited Recovery and Resilience:** Cyberfantastic systems are characterized by their innate agility, allowing for rapid recuperation after incidents. Such nimbleness minimizes the window of vulnerability and diminishes the repercussions of breaches. This inherent capability to rapidly bounce back from adversities naturally eases the pressure on continuous risk management.
4. **Autonomous Rectification and Adaptive Protocols:** Cyberfantastic systems come with an in-built ability: autonomous rectification. Upon detecting anomalies, they instinctively launch remedial actions, offsetting risks as they emerge, thereby attenuating the frequency of manual interventions.
5. **Strategic Resource Deployment:** A pivot to Cyberfantastic means organizations can be judicious with their resource allocation. Instead of committing heavily to perpetual risk oversight, energies can be channeled to fortify system adaptability. This approach not only ensures efficient resource utilization but also eases off the traditional risk management accelerator.

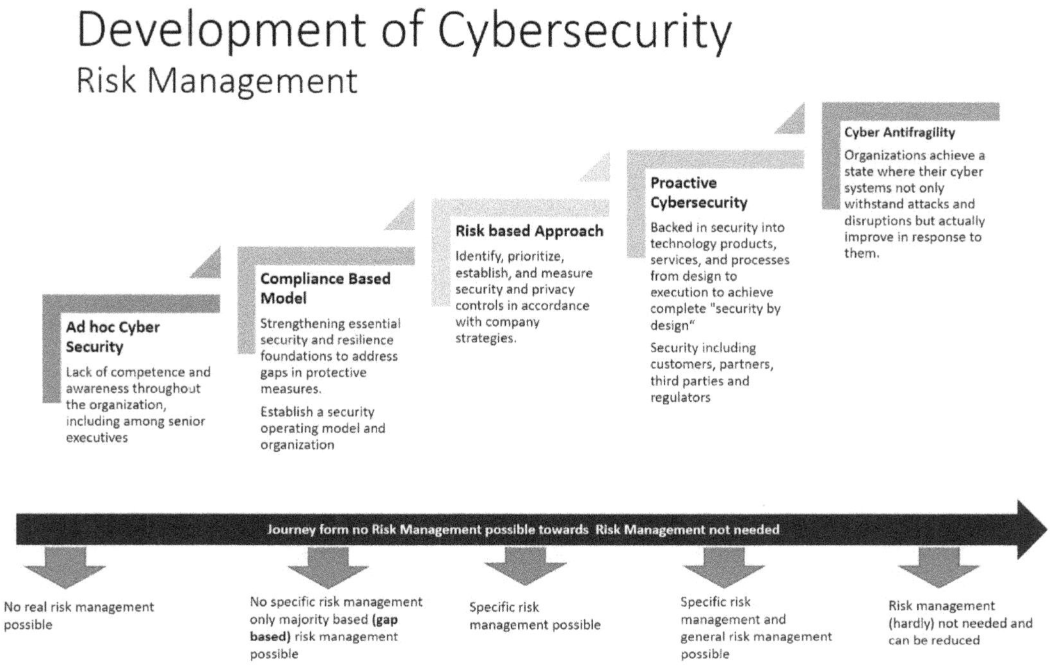

Figure 12.2 Risk management within the five steps (ad hoc, compliance, risk based, proactive, and Cyberfantastic) of cyber evolution of a company.

6. **Amplified Operational Efficiency:** Cyberfantastic inherently mandates operational process enhancement. By emphasizing dynamic, autonomous systems, organizations expedite incident responses and recovery, streamlining potential risk management and minimizing the need for elaborate risk handling protocols.

To encapsulate, the journey toward Cyberfantastic surpasses traditional risk management's confines. While risk management remains a foundational pillar, its modus operandi and emphasis are redefined. Cyberfantastic champions systems that evolve from passive defense mechanisms to proactive entities that grow stronger with challenges. The dividends are manifold: augmented robustness, resource optimization, and a renewed operational zenith, all harmoniously orchestrated in the Cyberfantastic continuum (Figure 12.2).

The Evolutionary Journey of Cybersecurity and Risk Management Paradigms

1. **Emergence – Stage 1:** In the early stages of the cybersecurity evolution, the landscape is characterized by chaos and uncertainty. Organizations find themselves in a vulnerable position due to the absence of defined strategies and a limited understanding of threats. Comprehensive risk management seems like an unattainable goal in this nascent phase.
2. **Foundational Compliance – Stage 2:** As the importance of security becomes evident, organizations transition to a compliance-driven approach. During this phase, risk management remains primarily reactive, with a focus on meeting regulatory requirements and addressing glaring security vulnerabilities. Strategies are often broad and basic, primarily aligned with industry standards.

3. **Tailored Risk Assessment – Stage 3:** The subsequent stage marks a sophisticated shift toward risk-centric security. Equipped with systematic risk assessments, organizations begin to develop and implement security measures specifically tailored to combat identified threats. Risk management becomes more targeted, moving beyond compliance-oriented practices to genuinely comprehend and address cyber risks. Automation begins to play a more prominent role in risk assessment, reducing the likelihood of judgment errors in qualitative risk management.
4. **Collaborative Security – Stage 4:** With cybersecurity threats growing in complexity and interconnectedness, a proactive and collaborative approach emerges. Organizations not only bolster their defenses but also actively engage with peers, industry groups, and even competitors to share intelligence and strategies. This collaborative spirit results in a fusion of specific and generalized risk management tactics, effectively countering multifaceted threats.
5. **Highpoint with Cyberfantastic – Stage 5:** At this stage, cybersecurity enters its most adaptive and automated phase. The Cyberfantastic paradigm envelops systems, empowering them not only to withstand threats but also to evolve and thrive in their aftermath. While risks still exist, the innate ability of systems to learn, adapt, and automate risk management processes diminishes the traditional imperative of manual risk assessment. Quantitative risk management benefits from reduced estimation errors as automated tools provide more accurate and dynamic risk assessments.

In Stage 5, automation becomes the cornerstone of risk management. Advanced machine learning (ML) algorithms continuously analyze and adapt to emerging threats in real time, reducing the potential for human judgment errors. Automated systems monitor and respond to incidents with precision, minimizing reaction time and mitigating potential damage. This level of automation allows organizations to manage risks proactively, ensuring that their cybersecurity defenses remain dynamic and adaptive.

The pivotal role of automation in risk management cannot be overstated. With limited resources at hand, organizations must direct their efforts effectively. Automation streamlines risk assessment, enabling organizations to allocate resources where they are needed most. It optimizes the utilization of cybersecurity personnel and technology, ensuring that the focus is on addressing real, data-driven risks rather than hypothetical scenarios. This strategic approach enhances the overall cybersecurity posture, making the best use of limited resources and ultimately strengthening cyber defenses.

EMBARKING ON THE PATH TO CYBERFANTASTIC: A PROGRESSIVE ODYSSEY

The journey to achieving optimal cybersecurity isn't merely about erecting barriers against threats; it's about evolving one's entire approach and infrastructure to not just weather cyber storms but flourish amid them. This progression can be visualized as a continuum, ranging from rudimentary measures to the possible highpoint of cybersecurity, termed here as "Cyberfantastic." It's important to note that while each stage builds upon the foundation of the previous ones, organizations do not need to wait for one stage to be completed before initiating activities from different stages. This approach fosters a proactive stance toward cybersecurity. Here's a step-by-step guide through this transformative journey:

Stage 1: Ad hoc security

The Beginning: Often, organizations start with a haphazard approach to security. It's about firefighting rather than fire prevention.

- **Assessment:** Begin by assessing the current cybersecurity landscape, identifying vulnerabilities through vulnerability scanning and penetration testing. Understand potential threats by analyzing historical attack patterns.
- **Awareness:** Enlighten all stakeholders, from top executives to front-line employees, on the indispensability of cybersecurity. Conduct cybersecurity awareness training sessions to ensure that everyone understands their role in maintaining security.
- **Foundational Measures:** Roll out essential security tools such as firewalls, antivirus software, intrusion detection systems, and swift incident response plans to address immediate threats. Implement basic access controls and password policies.
- **Time & Budget:** This stage typically spans a few months to a year, with a budget ranging from the thousands to tens of thousands. Investment is primarily in foundational security tools and training.

Stage 2: Compliance-based security

The Foundation: As maturity sets in, there's a recognition of the importance of aligning with industry standards and regulations.

- **Regulatory Alignment:** Ensure congruence with pertinent industry norms and regulatory requirements. Compliance with regulations such as NIS-2, GDPR, HIPAA, or industry-specific standards becomes a key focus. Establish a compliance team responsible for tracking and ensuring adherence.
- **Security Policies:** Craft cybersecurity policies that echo industry benchmarks and incorporate regulatory requirements. Policies should cover areas such as data protection, incident response, and access control. Document procedures for compliance auditing.
- **Security Training:** Constantly enlighten employees about the mutable cyber environment. Regularly conduct cybersecurity training sessions and awareness campaigns. Provide guidance on compliance-related responsibilities and reporting.
- **Time & Budget:** This phase might extend up to a year or more, costing anywhere from tens to hundreds of thousands. Investment includes compliance efforts, policy development, and ongoing training programs.

Stage 3: Risk-based security

The Deep Dive: Organizations now prioritize threats, moving beyond generic measures to more tailored interventions.

- **Risk Assessment:** Conduct a rigorous examination of risks, ranking them based on their potential repercussions to the organization. Collaborate with cybersecurity experts to perform comprehensive risk assessments, utilizing methodologies like for example FAIR (Factor Analysis of Information Risk).
- **Tailored Controls:** Design and implement security mechanisms specific to the highlighted risks. This might involve deploying intrusion detection systems, advanced access controls, data encryption, and application security measures tailored to identified vulnerabilities.

- **Continuous Monitoring:** Keep a vigilant eye on systems, monitoring them in real time for any deviations from the established security baselines. Implement Security Information and Event Management (SIEM) systems for centralized threat detection and incident response.
- **Time & Budget:** This stage requires a comprehensive 1–2 years of effort, with expenses potentially reaching a million or more. Investment includes risk assessments, security controls, and advanced monitoring solutions.

Stage 4: Proactive security

The Outreach: Cybersecurity isn't an insular activity. It involves building bridges, collaborating, and learning from the broader ecosystem.

- **Collaboration:** Forge alliances with stakeholders, including other organizations, industry groups, and even competitors. Establish threat sharing communities and participate in information sharing and analysis centers (ISACs) to pool threat intelligence. Collaborative threat hunting and incident response initiatives are essential.
- **Third-Party Risk Management:** Ensure that external associations and partners don't become the weak link in your cybersecurity chain. Conduct thorough risk assessments of third-party vendors and partners. Implement contractual requirements for security and compliance.
- **Advanced Threat Detection:** Upgrade to sophisticated threat discernment mechanisms such as ML-based anomaly detection, behavioral analytics, and threat hunting teams. Employ deception technologies to lure and identify attackers.
- **Time & Budget:** This stage requires a commitment of 2–3 years, with finances possibly touching the millions, depending on the scale and complexity of the organization's operations. Investment includes collaboration efforts, third-party risk assessments, and advanced threat detection capabilities.

Stage 5: Cyberfantastic

The Highpoint: This stage is where organizations don't merely survive challenges but thrive because of them.

- **Cultural Shift:** Instill a culture that perceives challenges as catalysts for evolution. Foster a cybersecurity-aware culture where every employee is a stakeholder in cybersecurity. Promote innovation, creativity, and continuous learning throughout the organization.
- **Autonomous Systems:** Deploy systems that can independently identify and neutralize threats. Implement advanced artificial intelligence (AI) and ML solutions that continuously learn from threats and adapt to evolving attack techniques. These systems can autonomously respond to threats in real time.
- **Antifragility Testing:** Intentionally simulate adversities to test and enhance system robustness. This involves subjecting systems to controlled cyberattacks to identify weaknesses, improve resilience, and train AI-based systems to respond effectively.
- **Time & Budget:** A long-haul commitment spanning several years is essential for this stage, with a fluid budget to accommodate pioneering solutions and the development of Cyberfantastic capabilities. Investment includes cultural initiatives, autonomous cybersecurity systems, and rigorous antifragility testing.

Reflecting back

This evolutionary path is a broad framework, with real-world scenarios influenced by myriad factors like the organization's inherent intricacies and the ever-shifting cyber threat landscape. Achieving Cyberfantastic isn't the end but a significant milestone in the ongoing marathon of cybersecurity. This commitment not only fortifies an organization's digital fortress but positions it to grow and innovate amid challenges, signaling its true metamorphosis in the cyber realm.

THE PHILOSOPHY OF CYBERFANTASTIC

In an era dominated by the complexities of the digital realm, the pursuit of a resilient and adaptable cybersecurity approach becomes paramount. The narrative of the fictional entity "NexTech Ventures" illuminates this journey, offering a roadmap to a digital landscape where resilience and adaptability reign supreme. Delve into this narrative that reshapes our comprehension of cybersecurity, weaving a vivid tapestry of innovation, strategy, and unwavering strength.

Chapter 1: The Digital Dawn of Resilience Amidst the vast expanse of digital frontiers, NexTech Ventures embarks on an enthralling quest toward the realization of Cyberfantastic. Their vision is a beacon of light, illuminating the path to crafting a dynamic digital environment where challenges not only fail to weaken but serve as catalysts for invigoration and advancement.

Chapter 2: Gathering the Digital Vanguard a resounding call reverberates throughout the digital domain, summoning a formidable ensemble of individuals. Within this coalition, security analysts stand as vigilant guardians, constantly on the lookout for potential threats and orchestrating swift countermeasures. Data alchemists and AI virtuosos delve deep into the realm of algorithms, ensuring that the system's cognitive capabilities intensify with each new challenge. Legal and compliance protectors vigilantly watch over the realm, ensuring that each technological advancement remains within the boundaries of digital regulations. Meanwhile, IT craftsmen meticulously shape the robust backbone of the digital fortress, ensuring its resilience and readiness for the future.

Chapter 3: Crafting the AI Beacon: Continuous Cognition the coalition taps into the profound intelligence of artificial minds, giving birth to a sentinel AI. This entity possesses the ability to absorb, learn, and predict from every byte of data, emerging as the cognitive core of the fortress.

Chapter 4: The Autonomous Knights Arise From the foggy mists of challenges emerges a revelation: autonomous systems. These systems, equipped with unmatched precision, counter threats autonomously, refining their prowess with each engagement.

Chapter 5: Glimpses of Tomorrow: The Predictive Oracle Harnessing vast data streams, NexTech Ventures forges a digital seer. Through Predictive Analytics, they transition from mere defense to foresight, preparing their defenses for unseen future threats.

Chapter 6: The Resilient Bastion: Unyielding & Self-Healing At the core of resilience, they craft an Immutable Infrastructure. Each digital component is imbued with Zero Trust, ensuring no invader gains traction, and the system rejuvenates faster after each challenge.

Chapter 7: Quantum Safeguards: Guardians of the Future in preparation for the looming quantum age, NexTech Ventures fortifies its haven with unparalleled shields. Quantum-defiant barriers and next-gen encryption methods are woven into the fabric, ensuring steadfast defense against the uncertainties of tomorrow.

Chapter 8: The Hybrid Ballet: Graceful, Unyielding Adaptations a mesmerizing dance of resilience unfolds as the Hybrid Cloud strategy takes center stage. Operations glide gracefully across platforms, ensuring fluidity and nimbleness, regardless of the digital weather.

Chapter 9: The Enigmatic Veil: Fortifying Privacy with a profound respect for the sanctity of privacy, NexTech Ventures employs a multilayered data protection strategy. Data transforms into enigmatic riddles, shielded by intricate layers, confounding even the most determined invaders.

Chapter 10: The Randomness Ruse: Unpredictability as Armor in a stroke of genius, NexTech Ventures introduces controlled chaos into its defense strategies. The infusion of randomness ensures that adversaries are perpetually perplexed, forever trailing in their pursuit.

Chapter 11: The Knowledge Nexus: The Power of Feedback The citadel's most radiant gem is unveiled: The Feedback-Driven Core. This vibrant heart distills insights from every digital skirmish, perpetually refining defenses, ensuring an ever-evolving state of readiness against adversity.

This narrative of NexTech Ventures embodies a philosophical journey that transcends the conventional paradigms of cybersecurity, leading to a realm where challenges become opportunities, and where resilience and adaptability prevail in the face of digital adversaries.

WRAP-UP: PIONEERING A NEW ERA OF CYBERSECURITY THROUGH CYBERFANTASTIC

The journey from vulnerability to resilience, culminating in the realm of Cyberfantastic, signifies a fundamental shift in the approach to safeguarding digital assets. This voyage has traversed through five distinct stages, each contributing to the development of systems that not only withstand adversity but thrive amidst it.

Starting with Ad Hoc Security in Stage 1, organizations often employ rudimentary measures and firefighting approaches. As maturity grows, the transition to compliance-based security in Stage 2 aligns with industry standards and regulations. Stage 3 witnesses a deep dive into risk-based security, where tailored interventions and continuous monitoring become the norm. In Stage 4, the emphasis shifts to proactive security, characterized by collaboration and advanced threat detection. Finally, in Stage 5, reaching the pinnacle of Cyberfantastic, where systems not only survive but evolve and thrive through adversities.

The transformative power of Cyberfantastic extends beyond technical innovations; it permeates the culture, reshaping how threats are perceived and responded to. Challenges are no longer viewed as obstacles, but rather as opportunities for growth and evolution.

Crucially, this journey redefines the concept of risk management. As systems become self-adapting and resilient, the need for traditional risk mitigation lessens. Incidents serve as valuable lessons, ingrained into the system's DNA, enabling autonomous responses to potential vulnerabilities. This realignment of resources and focus empowers organizations to proactively innovate while maintaining a robust security posture.

Society may well stand on the brink of a new era in cybersecurity, propelled by the concept of Cyberfantastic. It inspires a reimagination of strategies, the embrace of adaptability, and the forging ahead with a mindset that transcends the fear of threats. Although challenges lie ahead, the promise of systems that thrive through adversity beckons toward a future where the digital presence is not just secure but truly fantastic.

The benefits of Cyberfantastic extend to each element of the Parkerian Hexad (https://en.wikipedia.org/wiki/Parkerian_Hexad), an extension of the CIA Triad, as follows:

- **Confidentiality:** Cyberfantastic actively responds to threats, ensuring the confidentiality of sensitive information even in adversarial conditions. Systems continually adapt to protect data from breaches and unauthorized access.
- **Integrity:** Cyberfantastic safeguards data integrity by actively detecting and mitigating attempts at data tampering. Systems that learn from threats maintain the accuracy and trustworthiness of data.
- **Availability:** Cyberfantastic enhances availability, allowing systems to thrive under stressors and attacks, ensuring uninterrupted service for critical functions.
- **Possession (or Control):** Cyberfantastic strengthens possession and control by actively safeguarding resources, preventing unauthorized parties from seizing control during adverse events.
- **Authenticity:** Continuous identity verification is enhanced by Cyberfantastic, which dynamically adapts authentication mechanisms to counter evolving threats, maintaining trust and authenticity.
- **Utility:** Utility is significantly enhanced as systems remain functional and useful despite disruptions and attacks, enabling organizations to leverage digital resources without substantial interruptions.

In summary, the benefits of Cyberfantastic closely align with the values of the Parkerian Hexad. By actively responding to threats and challenges, Cyberfantastic enhances the security, reliability, and trustworthiness of digital systems, ultimately safeguarding information and resources in terms of confidentiality, integrity, availability, possession, authenticity, and utility.

A CALL FOR COLLABORATION: FOSTERING CYBERFANTASTIC IN AN ECOSYSTEM OF INNOVATION

In the relentless pursuit of Cyberfantastic, it's crucial to recognize that the journey thrives within an ecosystem of innovation and collaboration. The transformative paradigm of Cyberfantastic is less likely to be achieved by individual vendors or organizations alone; it flourishes when multiple stakeholders converge, fostering an environment of shared knowledge, expertise, and resources.

The power of ecosystems

In the digital age, where threats evolve rapidly and sophistication knows no bounds, the significance of a collaborative ecosystem cannot be overstated. The probability of achieving Cyberfantastic exponentially grows when multiple entities contribute their unique insights and capabilities.

Collective intelligence

Cybersecurity challenges are multifaceted, encompassing a diverse range of threats, vulnerabilities, and attack vectors. A collaborative ecosystem facilitates the pooling of collective intelligence. Security experts, researchers, and technologists from various domains create a rich tapestry of knowledge that is more adept at identifying, mitigating, and preventing threats.

Comprehensive defense

Cyberfantastic entails holistic resilience and adaptability. An ecosystem approach allows for the integration of diverse solutions and strategies. This inclusivity ensures that no aspect of cybersecurity is neglected, from threat detection and incident response to data protection and privacy preservation.

Faster innovation cycles

The digital landscape is in a perpetual state of evolution. An ecosystem fosters faster innovation cycles as ideas are shared, refined, and implemented collectively. This agility is essential to stay ahead of emerging threats and to continuously improve the Cyberfantastic model.

Leveraging resources

Cybersecurity is resource-intensive. By working within an ecosystem, organizations can pool resources, reducing the burden on individual entities. This shared resource pool enables smaller organizations and startups to access the tools and expertise needed to contribute effectively to the Cyberfantastic initiative.

The role of government and regulatory bodies

Government agencies and regulatory bodies play a vital role in shaping the cybersecurity landscape. They can facilitate collaboration, set standards, and provide a framework for information sharing within the ecosystem. Government-industry partnerships are instrumental in achieving Cyberfantastic on a broader scale.

International collaboration

Cyber threats are not constrained by borders. International collaboration within the ecosystem is essential to address global cybersecurity challenges. Information sharing, threat intelligence exchange, and coordinated responses are critical components of a robust Cyberfantastic ecosystem.

The path forward

As the journey toward Cyberfantastic unfolds, it's imperative to recognize the inherent strength of collaborative ecosystems. Individual efforts are commendable, but the collective power of an ecosystem amplifies the ability to transform the cybersecurity landscape. By fostering partnerships, sharing knowledge, and collectively innovating, the evolving digital terrain can be navigated with resilience, adaptability, and the promise of a Cyberfantastic future.

FINAL THOUGHTS: NAVIGATING THE CHALLENGES ON THE ROAD TO CYBERFANTASTIC

Embarking on the quest for Cyberfantastic, it is essential to acknowledge that this transformative journey is not without its challenges. The Cyberfantastic model offers a compelling approach to addressing the diverse array of cybersecurity threats and challenges, requiring thoughtful contemplation of several considerations.

Overcoming cultural resistance

One of the primary hurdles in the pursuit of Cyberfantastic is the potential resistance from traditional organizational cultures. Shifting the mindset from viewing disruptions as threats to embracing them as catalysts for growth can be met with resistance. Effective change management strategies and leadership commitment are pivotal in overcoming this challenge. Encouraging a culture of innovation and adaptability can pave the way for a smoother transition.

Navigating technical complexities

Integrating newer technologies and protocols, especially within legacy systems, can be a complex endeavor. The technical intricacies of the Cyberfantastic model may disrupt current operations. However, it is crucial to view these disruptions as opportunities for technological advancement. Thorough planning, meticulous testing, and a phased implementation approach can mitigate technical challenges.

Resource allocation

The journey to Cyberfantastic may necessitate significant investments in terms of time, financial resources, and personnel training. Organizations must be prepared for these resource commitments. Conducting a comprehensive cost-benefit analysis can provide a clearer understanding of the long-term advantages and the return on investment that Cyberfantastic offers. This analysis can help stakeholders make informed decisions and justify continued investment.

Leveraging emerging technologies

Emerging technologies play a pivotal role in realizing the Cyberfantastic vision. Blockchain technology, with its tamper-evident logs and transactions, enhances data integrity and authenticity. Edge computing decentralizes data processing, reducing vulnerabilities associated with centralized systems. Augmented Reality (AR) and Virtual Reality (VR) can revolutionize training, simulation, and threat modeling. Embracing these technologies as enablers of Cyberfantastic is essential for its successful implementation.

Financial considerations

The economic implications of transitioning to the Cyberfantastic paradigm should not be underestimated. While upfront costs may be substantial, the potential benefits in terms of reduced breaches, enhanced system uptime, and proactive threat management can yield significant long-term savings. Regularly evaluating the return on investment can help organizations stay aligned with the Cyberfantastic vision and justify ongoing investments.

Engaging stakeholders

Stakeholder engagement is critical throughout the Cyberfantastic journey. Top management's commitment is essential for resource allocation and setting the strategic direction. IT professionals, as frontline defenders, provide valuable insights for refining the approach and addressing technical challenges. End-users, often the target of cyber threats, can be transformed from potential vulnerabilities into assets through effective training and awareness campaigns.

Anticipating the future

In the ever-evolving digital landscape, it is imperative to anticipate future developments. The rise of quantum computing, with its potential for faster computation, introduces new opportunities and threats that Cyberfantastic strategies must adapt to. Global regulatory changes may result in increased cybersecurity standards and compliance requirements. As cyber threats evolve in sophistication, the principles of Cyberfantastic must remain dynamic and forward-thinking.

While the path to Cyberfantastic may present challenges and uncertainties, the potential rewards in terms of enhanced security, adaptability, and growth are unparalleled. Organizations embarking on this journey must remain adaptable, forward-thinking, and collaborative. Together, we can ensure that the digital realm becomes not only more secure but also more resilient and vibrant. The Cyberfantastic vision beckons us to embrace the future of cybersecurity with optimism and determination.

THE NEEDED CONDITIONS FOR CYBERFANTASTIC

In the evolution toward a Cyberfantastic framework, there are key foundational conditions that organizations must establish. This section aims to elucidate these conditions and the requisites for their implementation.

1. **Organizational Commitment**
   - **Contextual Background:** In the face of growing digital operations, stringent regulatory imperatives, and the historical context of security breaches, organizational acknowledgment of cybersecurity's pivotal role has never been more pressing.
   - **Necessary Provisions:** A top-tier, organization-wide endorsement for cybersecurity initiatives. This should manifest as dedicated budgetary allocations, an overarching cultural emphasis on security awareness, and delineated roles and responsibilities within the security realm.
2. **Risk Assessment and Understanding**
   - **Contextual Background:** Comprehensive risk evaluations to unearth vulnerabilities and potential threats are paramount.
   - **Necessary Provisions:** Systematic and recurring risk assessments. Additionally, a dynamic model to rank risks according to their consequential impacts on organizational operations.
3. **Skilled Workforce**
   - **Contextual Background:** The cybersecurity landscape is in perpetual flux, demanding a workforce adept in contemporary and prospective threat vectors.
   - **Necessary Provisions:** A sustained commitment to workforce development through training initiatives, strategic recruitment of subject matter experts, and fostering a culture of continuous learning.
4. **Technological Infrastructure**
   - **Contextual Background:** The backbone of any Cyberfantastic endeavor is an advanced technological infrastructure.
   - **Necessary Provisions:** Infrastructure that is periodically assessed for security robustness and is inherently scalable. Legacy systems should be assessed for compatibility or considered for phase-out.
5. **Regulatory Compliance**
   - **Contextual Background:** Different industries and geographic locales have specific cybersecurity mandates.

- **Necessary Provisions:** Staying abreast of pertinent regulations and instituting internal mechanisms for compliance. This should include scheduled audits, reporting frameworks, and deployment of necessary controls.

6. **Risk Tolerance and Business Continuity**
   - **Contextual Background:** Every organization possesses a unique risk tolerance threshold.
   - **Necessary Provisions:** Clearly articulated risk appetite statements and comprehensive business continuity designs that ensure operational resilience in adverse cyber events.
7. **Innovation and Adaptation**
   - **Contextual Background:** The cybersecurity ecosystem is enriched by innovative methodologies and technologies.
   - **Necessary Provisions:** Regular horizon scanning for emerging cybersecurity methodologies and rapid integration of validated technologies.
8. **Collaboration and Information Sharing**
   - **Contextual Background:** Collective defense through collaboration amplifies organizational defenses.
   - **Necessary Provisions:** Actively participating in cybersecurity coalitions, fostering intelligence-sharing agreements, and transparent threat data sharing.
9. **Strategic Planning**
   - **Contextual Background:** Strategic foresight ensures long-term alignment of cybersecurity endeavors with organizational trajectories.
   - **Necessary Provisions:** Drafting and periodically revising a comprehensive cybersecurity strategy that aligns with, and supports, broader organizational visions.
10. **Testing and Incident Response:**
    - **Contextual Background:** Proactive testing is a precursor to identifying chinks in the armor.
    - **Necessary Provisions:** Routine penetration tests, vulnerability assessments, and a well-documented and practiced incident response strategy.

Transitioning to a Cyberfantastic framework necessitates a holistic approach, anchored in the aforementioned conditions. The journey to Cyberfantastic isn't about reaching a terminus but is about constant evolution, adaptation, and, above all, preparedness for the unforeseeable.

IMPLEMENTATION ROADMAP FOR ACHIEVING CYBER ANTIFRAGILITY

Achieving Cyber Antifragility requires a structured and well-defined implementation plan that encompasses strategic actions, compliance and legal oversight, capacity building, performance metrics, and collaboration within the cybersecurity ecosystem. This comprehensive roadmap serves as a guide for organizations aspiring to attain Cyber Antifragility, offering a detailed and strategic approach.

Strategic action points

1. **Formation of a Dedicated Task Force**
   Objective: Assemble a multidisciplinary team of experts to champion the antifragility vision.

Actions

- Onboard specialists such as Security Analysts for continuous monitoring, data scientists for ML integration, legal experts for regulatory compliance, and IT professionals for infrastructure innovation.
- Host monthly alignment sessions to ensure strategic cohesion.

2. **Integration of AI & Adaptive Learning**
   Objective: Create a dynamic defense system that evolves continually in response to emerging threats.
   Actions
   - Collaborate with data professionals to integrate suitable ML platforms.
   - Establish continuous learning protocols and strengthen computational facilities for real-time analysis.
3. **Advancement to Autonomous Mechanisms**
   Objective: Enhance operational efficiency by minimizing manual interventions in incident response.
   Actions
   - Develop Autonomous Incident Response systems capable of swift, automated threat mitigation.
   - Continuously refine autonomy protocols based on real-world experiences and threat intelligence.
4. **Predictive Analytics Infrastructure**
   Objective: Shift from reactive cybersecurity to proactive defense by identifying vulnerabilities before they are exploited.
   Actions
   - Leverage data analytics tools to detect vulnerabilities in their nascent stages.
   - Collaborate with security analysts to validate predictions and develop proactive mitigation strategies.
5. **Adoption of Immutable Infrastructure & Zero Trust Approach**
   Objective: Ensure data integrity even in the face of potential breaches by embracing immutable digital setups and Zero Trust principles.
   Actions
   - Transition to an immutable digital infrastructure that prevents unauthorized alterations.
   - Regularly review and update access permissions in alignment with Zero Trust philosophies.
6. **Integration of Quantum-Resistant Measures**
   Objective: Stay ahead of emerging computational threats by incorporating quantum-resistant cryptographic solutions.
   Actions
   - Implement quantum-resistant cryptographic measures.
   - Explore the potential of Quantum Key Distribution (QKD) for enhanced encryption.
7. **Assurance of Hybrid Cloud Resilience**
   Objective: Sustain operations irrespective of challenges by adopting a Hybrid Cloud strategy.
   Actions
   - Develop a comprehensive strategy that emphasizes Hybrid Cloud adoption.
   - Define criteria for automatic workload transitions between localized and cloud environments.

8. **Enhancement of Data Privacy with Advanced Routing**
   Objective: Strengthen the security of data in transit through advanced routing techniques and layered encryption.
   Actions
   - Integrate advanced routing methods to enhance encryption during data transmission.
   - Periodically update routing strategies to maximize data privacy and security.
9. **Infusion of Controlled Randomness**
   Objective: Confound and counteract predictable threat patterns by introducing controlled randomness in security operations.
   Actions
   - Incorporate controlled randomness in various security processes.
   - Regularly assess the effectiveness of randomized security measures in thwarting threats.
10. **Cultivation of Feedback-Driven Evolution**
    Objective: Leverage insights gained from cybersecurity incidents for continuous improvement.
    Actions
    - Establish feedback channels after every significant cybersecurity event.
    - Conduct monthly sessions to distill learnings and refine cybersecurity protocols.

Compliance and legal oversight

Ensure all cybersecurity initiatives remain compliant with relevant regulations and legal frameworks. Engage the Legal and Compliance team on an ongoing basis and prioritize bi-annual audits to maintain standards and transparency.

Capacity building and knowledge transfer

Position the cybersecurity team at the forefront of advancements in the field. Organize quarterly training sessions, encourage participation in external events, and foster knowledge exchange among team members.

Performance metrics and review

Dedicate efforts to continuous improvement by establishing clear key performance indicators (KPIs) such as breach prevention rates and system reliability. Conduct exhaustive annual evaluations to refine the overarching cybersecurity strategy.

Collaborative ecosystem

Recognize that achieving Cyber Antifragility is more attainable through collaboration within the broader cybersecurity community rather than in isolation. Engage in knowledge sharing, collaborative research, and partnerships with other organizations, academics, and experts to collectively strengthen the cybersecurity landscape.

By following this comprehensive implementation roadmap, organizations can navigate the dynamic cybersecurity landscape and emerge more resilient, adaptable, and antifragile in the face of evolving threats and challenges.

DECODING CYBERFANTASTIC: A COMPREHENSIVE CYBERSECURITY APPROACH

The acronym CYBERFANTASTIC represents a robust cybersecurity philosophy that integrates various attributes to craft a system that is Yielding, Bolstering, Evolving, Reinforcing, Flourishing, Adaptive, Navigable, Tenacious, Alert, Secure, Trustworthy, Innovative, and Complete. The adjective "fantastic" resonates with qualities of excellence and exceptionality, aptly describing a forward-thinking cybersecurity system.

- **C for Cyber:** Denotes the focus on the digital domain of cybersecurity.
- **Y for Yielding:** An adaptable system that demonstrates flexibility without compromising its integrity.
- **B for Bolstering:** Embodies the idea of continuously enhancing and solidifying defenses.
- **E for Evolving:** Underscores the system's ability to modify and adapt in a dynamic threat landscape.
- **R for Reinforcing:** Indicates a commitment to routinely upgrade and strengthen the defenses.
- **F for Flourishing:** Not merely about surviving threats but also thriving in their presence, demonstrating growth and progression.
- **A for Adaptive:** Emphasizing agility and swift adaptation in the face of rapidly changing cyber threats.
- **N for Navigable:** Suggests a user-friendly system that's easily maneuverable while maintaining robust defenses.
- **T for Tenacious:** Represents unwavering persistence and resilience against potential breaches.
- **A for Alert:** A system that's perpetually vigilant, ready to detect and combat threats proactively.
- **S for Secure:** Demonstrates a foundational attribute of unwavering security and protection.
- **T for Trustworthy:** Reiterates the system's reliability and its promise to guard against adversities.
- **I for Innovative:** Celebrates the ongoing commitment to incorporate novel solutions and state-of-the-art technologies.
- **C for Complete:** Advocates for a comprehensive approach that covers every facet of cybersecurity, leaving no stone unturned.

THE UPDATE

Observations: Enhancing the Cyberfantastic vision

Introduction to observations

In the ever-evolving landscape of cybersecurity, the advent of the Cyberfantastic paradigm marks a significant shift in how we conceptualize and implement digital defenses. While the foundational white paper sets the stage for this transformative approach, it is essential to delve deeper and expand upon these ideas to fully appreciate their impact and applicability.

The "Observations" section is an extension of the original white paper, designed to address critical feedback, incorporate practical insights, and explore the multifaceted implications of the Cyberfantastic concept. This section aims to bridge the gap between theoretical frameworks and real-world application, offering a comprehensive understanding of how Cyberfantastic principles can be realistically and effectively integrated into contemporary cybersecurity strategies.

Each segment of this extension delves into specific aspects of the Cyberfantastic model, ranging from the incorporation of real-world case studies that demonstrate these concepts in action, to addressing the practical challenges and legal considerations inherent in adopting such a forward-thinking approach. The goal is to provide a more rounded, pragmatic perspective, ensuring that the Cyberfantastic model is not only visionary but also grounded in the realities of today's digital world.

As we navigate through these observations, we will explore the nuances of building a resilient, adaptive, and proactive cybersecurity framework. This journey will take us through the complexities of network security, the human elements of building a Cyberfantastic team, and the future-oriented thinking required to stay ahead in a rapidly advancing technological era.

By expanding upon the original ideas presented in the white paper, the "Observations" section seeks to enrich the dialogue around Cyberfantastic, offering a roadmap for organizations and individuals eager to embrace this innovative approach. In doing so, we aim to contribute to the development of a more resilient, adaptable, and forward-thinking cybersecurity ecosystem.

1. **Implementing Real-World Case Studies**
   - **Possible Strategic Initiatives:** Develop a framework for integrating case studies into the Cyberfantastic model. Collaborate with diverse industries to document and analyze their cybersecurity challenges and solutions, focusing on the application of Cyberfantastic principles.
2. **Overcoming Practical Challenges**
   - **Possible Strategic Initiatives:** Identify common implementation challenges and develop a set of best practices and solutions. This includes creating training programs, resource allocation strategies, and methods to foster organizational buy-in for Cyberfantastic concepts.
3. **Ensuring Adaptability with Evolving Technology**
   - **Possible Strategic Initiatives:** Establish continuous learning programs and technology watch groups within organizations. Develop a process for regularly updating Cyberfantastic strategies in line with emerging technologies and threats.
4. **Integrating Antifragility and Continuous Testing**
   - **Possible Strategic Initiatives:** Design and implement a continuous testing framework using concepts like Chaos Monkey, Purple Team, and BAS. Develop protocols for automated responses and system adaptations based on test results.
5. **Establishing Controls for Chaos Engineering**
   - **Possible Strategic Initiatives:** Create guidelines and protocols for safely implementing chaos engineering. This includes the strategic use of honeypots and honeytokens and establishing boundaries and safeguards for these practices.
6. **Extending Antifragility to Third Parties**
   - **Possible Strategic Initiatives:** Develop guidelines for third-party partnerships that align with Cyberfantastic principles. Create assessment tools for evaluating and enhancing the cybersecurity resilience of third-party services.
7. **Cultivating a Cyberfantastic Team**
   - **Possible Strategic Initiatives:** Design a comprehensive training and development program focused on Cyberfantastic principles. Foster a culture that values adaptability, proactivity, and continuous learning in cybersecurity.

8. **Aligning with CMMI Maturity Model**
   - **Possible Strategic Initiatives:** Conduct an analysis to align Cyberfantastic stages with the Capability Maturity Model Integration (CMMI) model. Develop a roadmap for organizations to transition from traditional CMMI levels to the advanced stages of Cyberfantastic.
9. **Balancing Fragility, Robustness, and Compliance**
   - **Possible Strategic Initiatives:** Create a framework for balancing the embrace of vulnerabilities with compliance requirements. Develop strategies for innovative cybersecurity approaches within the confines of regulatory frameworks.
10. **Prioritizing Quantum Resistance**
    - **Possible Strategic Initiatives:** Initiate projects focused on quantum-resistant cybersecurity technologies. Include quantum resistance as a core component of the Cyberfantastic model, ensuring future-readiness.
11. **Analyzing Cost and Accessibility**
    - **Possible Strategic Initiatives:** Conduct a cost-benefit analysis of implementing Cyberfantastic strategies. Develop scalable and adaptable models of Cyberfantastic that can be implemented by organizations of various sizes and resources.
12. **Managing Network Complexity and Security**
    - **Possible Strategic Initiatives:** Develop protocols for handling the complexities of randomized routing and traffic. Create monitoring and response strategies that can adapt to the increased complexity while ensuring security.
13. **Preparing for Future States**
    - **Possible Strategic Initiatives:** Define and prepare for concepts like Cyber Reciprocity, Symbiosis, and Resonance. Develop a forward-looking strategy that positions Cyberfantastic as a foundation for these future cybersecurity paradigms.
14. **Applying Cyberfantastic in OT**
    - **Possible Strategic Initiatives:** Tailor Cyberfantastic principles for the unique requirements of Operational Technology. Develop specific strategies and protocols for enhancing cybersecurity resilience in industrial and infrastructure settings.
15. **Safeguarding against Adaptable Threats**
    - **Possible Strategic Initiatives:** Develop strategies for identifying and mitigating threats that specifically target the adaptable nature of Cyberfantastic systems. Focus on continuous innovation and vigilance in threat detection and response.
16. **Navigating Legal Considerations**
    - **Possible Strategic Initiatives:** Research and understand the legal implications of proactive cybersecurity measures in various jurisdictions. Develop a legal and ethical framework for implementing Cyberfantastic strategies, especially those involving proactive defense measures.

CYBER CHUCKLES: IS CYBER FANTASTIC STILL FANTASTIC?

Welcome aboard the Cyberfantastic voyage, where the cybersecurity landscape is not just a battleground but a playground for innovation! Buckle up as we embark on this rollercoaster ride from the ad hoc security mosh pits to the symphonic harmonies of Cyberfantastic.

From zero to hero: The Cyberfantastic evolution

Imagine starting in a world where "security" means telling your dog to watch the computer. That's your Stage 1, folks – charming but, let's face it, as effective as a chocolate teapot.

Then, as you evolve, you start actually following those pesky compliance rules (Stage 2), because apparently, "I didn't know" doesn't cut it in court.

Cruise along to Stage 3, where you start playing chess with your cyber foes, predicting moves, and actually making strategic plays based on risk. By Stage 4, you're not just playing the game; you're rewriting the rules, collaborating like the Avengers, and integrating third-party Jedi tricks.

Finally, welcome to Cyberfantastic (Stage 5), where your systems not only bounce back from attacks but do a little victory dance too. Here, vulnerabilities are just opportunities in disguise, and your digital infrastructure is more adaptable than a chameleon in a bag of Skittles.

Challenges: The plot twists

But oh, dear traveler, the path to Cyberfantastic is fraught with plot twists. From cultural resistance akin to introducing your grandma to Snapchat, to the Herculean task of integrating new tech without causing a digital apocalypse. And let's not forget about resource allocation – because apparently, money doesn't grow on digital trees.

The heroes of the tale: Implementing Cyberfantastic

In this epic, our heroes embark on strategic initiatives, wielding the mighty tools of AI and adaptive learning like digital Excaliburs. They harness the power of quantum resistance as if gearing up for a cyber Valhalla and embrace the chaos of randomness to keep foes on their toes.

Behold the creation of an autonomous, self-healing digital realm where even the smallest code snippet stands tall in the face of adversity. Imagine a cybersecurity dojo where every challenge is an opportunity to level up, and feedback loops are the secret potions for continuous improvement.

The moral of the story

In the end, dear reader, Cyberfantastic isn't just about having an impenetrable digital fortress. It's about embracing the journey, from the humble beginnings of ad hoc measures to the lofty heights of a system that thrives on challenge. It's a reminder that in the digital realm, being fantastic isn't just about surviving; it's about thriving, evolving, and occasionally, doing the impossible.

So, here's to the Cyberfantastic adventure – a tale of resilience, adaptability, and the undying spirit of innovation. May your systems grow stronger with every challenge, and may your digital future be as fantastic as the journey itself.

Chapter 13

Unraveling complexity

The silent adversary in cybersecurity

Spoiler Alert!

- Complexity: The arch-enemy of cybersecurity you never knew you had.
- This chapter dives into the hidden dangers of a complex cybersecurity landscape.
- It all starts with the ever-growing number of technologies, human error, and a never-ending stream of regulations.
- Imagine a tangled mess of wires and red tape – that's cybersecurity complexity in a nutshell.
- But fear not! This chapter also offers tips on how to fight back and simplify your security strategy (because sometimes, less really is more).

INTRODUCTION: UNVEILING THE SILENT ADVERSARY

The vast, expanding digital universe, where data streams intersect at the speed of thought and connections form the backbone of civilization, harbors a pervasive yet silent adversary. This adversary announces itself not with alarms or visible warnings but through its intricate weave into the digital fabric, unseen and often unrecognized. Complexity, this silent adversary, permeates the cybersecurity landscape, subtly influencing, disrupting, and frequently dictating the outcomes of cybersecurity efforts.

The digital age, characterized by unparalleled technological progress, has fostered an era of connectivity and innovation. However, each technological breakthrough, every new device linked to the internet, and all code developed add layers to an already labyrinthine cybersecurity landscape. The complexity in cybersecurity stems not just from technological growth but from the fundamental nature of our interconnected existence. It shapes the tactics of defenders and the strategies of attackers.

This narrative embarks on an analytical journey to the core of cybersecurity complexity. Its goal is to dissect its layers, trace its origins, and reveal its subtle impacts on the digital security battlefield. Complexity in cybersecurity emerges as a complex phenomenon, encompassing the technical intricacies of systems and networks, the human dynamics influencing behavior and decisions, and the regulatory frameworks governing practices. It presents both a challenge to surmount and a reality to acknowledge.

The exploration aims to illuminate the genesis of complexity within cybersecurity, examining its evolution and origins. It investigates complexity's dual role as both a catalyst for threats and an obstacle to effective defense mechanisms. Through examining complexity, the narrative scrutinizes its impact on cybersecurity defenses and discusses strategies for its mitigation. Concrete case studies will offer insights into combating complexity, providing lessons from cybersecurity's front lines.

DOI: 10.1201/9781003562184-18

Facing the silent adversary of complexity demands innovation, adaptation, and a drive toward simplicity. The exploration of cybersecurity complexity isn't merely academic; it's crucial for safeguarding the resilience and integrity of the digital domain. Navigating this complex landscape requires vigilance, innovation, and a commitment to deepening understanding of complexity's nuances and devising effective mitigation strategies.

This introduction invites readers on a comprehensive exploration of complexity in cybersecurity, aiming to demystify the intricate challenges that defenders face and embolden adversaries in the digital realm.

THE GENESIS OF COMPLEXITY IN CYBERSECURITY

The inception of complexity within the realm of cybersecurity tells a story not merely of technological advancement but of ambition, innovation, and the unintended consequences that accompany the march toward a digitized existence. This tale begins as humanity's technological aspirations birth an interconnected world, bringing with it not just the promise of boundless connectivity but also a complex web of interdependencies and vulnerabilities previously unimagined.

TECHNOLOGICAL EVOLUTION

The march of technology, from the foundational development of the internet to the advent of cloud computing and the proliferation of IoT devices, has been characterized by a relentless push toward greater connectivity and functionality. Each leap forward, while undoubtedly advancing the cause of innovation, has concurrently introduced new layers of complexity into the cybersecurity domain. This complexity manifests in the expanding surface area for potential cyberattacks, the increasing difficulty of securing diverse and dispersed digital assets, and the heightened challenge of protecting data across varied and often incompatible platforms.

HUMAN FACTORS

Beyond the technological landscape, the human element introduces its brand of complexity to cybersecurity. User behavior, marked by its unpredictability and diversity, adds a variable layer to cybersecurity challenges. From the end user susceptible to sophisticated phishing attacks to the system administrator managing sprawling network architectures, human actions and decisions weave additional threads into the complex tapestry of cybersecurity. Organizational structures further compound this complexity, as policies, procedures, and cultural dynamics influence the effectiveness of cybersecurity measures.

REGULATORY AND COMPLIANCE LANDSCAPE

The evolving landscape of laws, regulations, and standards introduces yet another dimension of complexity. Organizations find themselves navigating a minefield of compliance requirements that vary not just by industry but across geographical boundaries. The task of aligning cybersecurity practices with a myriad of regulatory frameworks demands a delicate balancing act, as compliance adds layers of procedural and technical complexity to an already challenging environment.

The genesis of complexity in cybersecurity is thus a multifaceted phenomenon, born of the interplay between technological progression, human behavior, and regulatory demands. This complexity, while fostering innovation and driving the expansion of the digital frontier, simultaneously engenders vulnerabilities and presents obstacles to effective cybersecurity.

In recognizing the origins of complexity, one gains insight into the foundational challenges that underpin the cybersecurity field. Understanding these roots is crucial for developing strategies that address not just the symptoms but the core issues that complexity introduces into the cybersecurity landscape.

COMPLEXITY AS A CATALYST FOR THREATS AND DEFENSE COMPLICATIONS

Complexity in the digital landscape serves a dual role that is paradoxical yet intrinsic to the nature of cybersecurity. On one hand, it drives technological innovation and connectivity, fostering advancements that have reshaped the world. On the other, it breeds vulnerabilities and chaos, offering cyber adversaries a fertile ground for exploitation. This section delves into the multifaceted nature of complexity, examining its influence on the amplification of cyber threats and the intricate challenges it presents to defense mechanisms.

Amplification of cyber threats

Complexity inherently increases the number of attack vectors available to cyber adversaries. The expansion of digital infrastructure, the integration of new technologies, and the increasing reliance on interconnected systems provide a broader landscape for potential attacks. Each new device, application, or network connection introduces possible entry points for attackers, multiplying the opportunities for exploitation. This diversification of attack vectors complicates the task of securing digital assets, as defenders must anticipate and protect against a continually evolving array of threats.

Moreover, the predictive models traditionally employed to anticipate cyberattacks struggle to adapt to the intricate web of modern digital interactions. Traditional threat models, designed for less complex environments, often fall short in accurately predicting attacks within highly interconnected systems. The unpredictability introduced by complexity makes it challenging for cybersecurity professionals to identify potential threats proactively, often relegating them to a reactive stance.

Adversarial adaptation

Attackers exploit the complexity of digital systems to their advantage, crafting attacks that are as intricate and elusive as the environments they target. By leveraging the convoluted nature of modern networks and systems, adversaries can devise sophisticated strategies that bypass traditional security measures. The obscurity and ambiguity inherent in complex systems provide cover for malicious activities, allowing attackers to operate undetected for extended periods. This exploitation of complexity not only facilitates the initial breach but also complicates the process of identifying and mitigating attacks, extending the potential damage inflicted.

DEFENSE MECHANISMS: THE DOUBLE-EDGED SWORD OF COMPLEXITY

While complexity poses significant challenges to cybersecurity efforts, it also impacts the defense mechanisms employed to protect digital assets. The deployment of numerous security solutions, intended to safeguard against a wide range of threats, often results in a security solution overload. This accumulation of security tools and measures can lead to increased complexity within the cybersecurity architecture itself, potentially introducing gaps in defense and making it more difficult to manage and coordinate effective responses.

Operational challenges emerge as organizations struggle to manage their increasingly complex cybersecurity architectures. The operational overhead associated with maintaining a multitude of security solutions includes not only financial costs but also the burden on human resources. Cybersecurity teams may find themselves overwhelmed by the task of configuring, updating, and monitoring an extensive array of security measures, detracting from their ability to focus on strategic security initiatives.

Furthermore, the concept of "security through obscurity" – relying on the complexity of systems to deter attackers – reveals its limitations within a complex ecosystem. While obfuscating system details can provide a temporary barrier to attackers, it also impedes the organization's ability to maintain clear visibility into its security posture. This lack of clarity and simplicity can obscure vulnerabilities, making it more challenging to identify and address security gaps effectively.

In navigating the intricate world of cybersecurity, recognizing the dual nature of complexity is crucial. It demands a nuanced approach that acknowledges complexity's role in both fostering innovation and introducing vulnerabilities. As the narrative progresses, strategies for mitigating the impact of complexity on cybersecurity will be explored, offering insights into maintaining resilience in the face of this silent adversary.

STRATEGIES FOR MITIGATING COMPLEXITY

Tackling the intricate challenges posed by complexity in cybersecurity requires a deliberate and methodical approach. It calls for strategic insight, a commitment to design simplicity, and a holistic view of security practices. This section outlines actionable strategies aimed at reducing the impact of complexity on cybersecurity defenses, enhancing the effectiveness of security measures, and streamlining cybersecurity management.

- **Principle of Least Privilege:** Fundamental to simplifying cybersecurity complexity is the rigorous application of the principle of least privilege across all systems and networks. By limiting users' access rights to only what is strictly necessary for their job functions, organizations can significantly reduce the attack surface, making it harder for attackers to exploit elevated privileges. This approach not only tightens security but also reduces complexity by minimizing the number of variables and interactions that need to be managed.
- **Unified Security Architecture:** The proliferation of disparate security tools and platforms can exacerbate complexity, leading to inefficiencies and potential security gaps. By adopting a unified security architecture, organizations can integrate various security tools into a cohesive, streamlined framework. This integration facilitates centralized management, improves visibility across the security landscape, and enhances the ability to detect and respond to threats more effectively. A unified approach reduces operational overhead and fosters a more manageable and coherent cybersecurity posture.

- **Cultural Shift Toward Security:** Mitigating complexity is not solely a technical endeavor but also a cultural one. Fostering a culture of security awareness and simplicity across the organization encourages stakeholders at all levels to prioritize security and embrace practices that minimize complexity. This cultural shift involves regular training, clear communication of security policies, and the promotion of security best practices that emphasize straightforward, effective solutions. Cultivating a security-aware culture empowers individuals to make informed decisions that contribute to reducing the overall complexity of the cybersecurity environment.

CASE STUDIES: LEARNING FROM COMPLEXITY

Understanding the practical implications of complexity in cybersecurity can be enhanced through the examination of real-world case studies. These examples offer insights into the tangible impacts of complexity on cybersecurity incidents, providing lessons from both successful complexity reduction initiatives and instances where complexity contributed to failures.

- **Success in Reducing Complexity:** An in-depth analysis of an organization that effectively managed to reduce its cybersecurity complexity provides valuable insights into successful strategies. This case study highlights the measures taken, such as streamlining security tools, implementing comprehensive access controls, and fostering a culture of security awareness. The positive outcomes, including improved efficiency, enhanced security posture, and reduced operational costs, underscore the benefits of actively addressing complexity.
- **Failure Exacerbated by Complexity:** Conversely, examining a case where complexity played a significant role in a cybersecurity failure reveals critical lessons. This analysis delves into the factors that contributed to the breach, such as overly complex security architectures, lack of visibility into the security landscape, and inadequate response mechanisms. The case study draws attention to the importance of simplicity, effective management, and the need for a unified security approach in preventing similar failures.

THE FUTURE LANDSCAPE: COMPLEXITY AND CYBERSECURITY EVOLUTION

As the digital ecosystem continues to evolve, so too does the relationship between complexity and cybersecurity. Emerging technologies promise to bring new capabilities and efficiencies but also introduce additional layers of complexity.

- **Emerging Technologies and New Complexities:** The advent of artificial intelligence, blockchain, and quantum computing is set to revolutionize the cybersecurity landscape. While these technologies offer potential breakthroughs in security capabilities, they also present new challenges and complexities that organizations must navigate. Understanding and preparing for the implications of these technologies is crucial for managing future cybersecurity complexity.
- **Adapting to Complexity:** The enduring challenge of complexity in cybersecurity necessitates ongoing adaptation and innovation. Cybersecurity professionals and organizations must remain agile, continuously updating their knowledge and

strategies to keep pace with the evolving digital environment. Embracing a mindset of continuous learning, leveraging emerging technologies judiciously, and maintaining a focus on simplicity will be key to thriving in the complex cybersecurity landscape of the future.

CONCLUSION

Navigating the multifaceted domain of cybersecurity complexity has revealed critical insights, strategies, and a compelling call to action. Facing the future requires vigilance, adaptability, and a dedication to embracing simplicity amid complexity. This exploration serves as a guide through the challenges of cybersecurity complexity, offering a deeper understanding of its implications and the tools for effective mitigation. By adopting a multifaceted approach to complexity, the cybersecurity community can navigate the evolving landscape with resilience, ensuring the security of our interconnected digital world.

CYBER CHUCKLES: SIMPLIFYING THE SECURITY SPAGHETTI

Welcome to the spaghetti bowl of cybersecurity, where more layers don't mean more flavor but more confusion. Imagine trying to untangle actual spaghetti with a fork made of code and policy documents – welcome to our world!

The complexity conundrum

Picture cybersecurity as a game of Jenga, but every block is a different technology or regulation. Pull the wrong one, and instead of a gentle tumble, you get a digital apocalypse. Our job? To keep the tower standing while somehow removing half the pieces without anyone noticing.

Decoding the tangle

Translating cybersecurity complexity is like trying to read an ancient scroll with instructions to save the world, but half of it's in emojis. "Implement robust firewall protocols" translates to "Put up a digital 'Keep Out' sign," and "Ensure end-to-end encryption" becomes "Make sure no one's eavesdropping on our digital gossip."

Streamlining with style

Cutting through the complexity doesn't require a machete but perhaps a scalpel. Or, in tech terms, turning off the 37 alarm systems we accidentally installed on the same door. It's about making "simplifying the security strategy" sound less like "Let's just hope for the best" and more like "We're going minimalist – only the essentials."

From chaos to clarity

Who knew that reducing cybersecurity complexity could be like organizing a cluttered garage? Suddenly, finding your security tools doesn't require a map and a treasure hunt. "Minimizing potential threat vectors" is no longer a game of Where's Waldo but a clear, open field where you can see threats coming from a mile away.

The hero's journey: From complexity to simplification

Our quest through the labyrinth of cybersecurity complexity is epic, filled with dragons (malware), trolls (hackers), and the occasional wizard (IT guru). By the end, we don't just emerge victorious with a simplified security strategy; we've also somehow managed to make the entire journey seem like a walk in the park.

In conclusion: Keep it simple, cyber soldier

As we wrap up this roller coaster ride through cybersecurity's complexity, remember: simplicity is your Excalibur. Cutting through the dense forest of technological and regulatory overgrowth might seem daunting, but with a clear head and a steady hand, it's possible. So, the next time you're faced with a "multi-dimensional security assessment," know it means "Let's actually figure out what's going on."

And so, armed with our newfound knowledge and a slightly odd collection of metaphors, we march forward, ready to turn the spaghetti bowl of cybersecurity into a neatly organized bento box. To simplification – and beyond!

Chapter 14

Navigating cybersecurity with precision

Spoiler Alert!

- This chapter ditches the compass and introduces KPIs (Key Performance Indicators) as your guide through the cybersecurity wilderness.
- Think of KPIs as measurable goals that help you understand how well your security efforts are working.
- Learn how to pick the right KPIs (like choosing the perfect hiking trail – some are scenic, others challenging).
- This chapter goes beyond picking KPIs – it shows you how to use them to improve your entire security posture (think of it as packing the right gear for your cybersecurity trek).
- Get ready for real-world examples: a successful bank leveraging KPIs and a healthcare provider learning from its mistakes (because everyone can benefit from a good cautionary tale).

INTRODUCTION

In the vast and ever-changing landscape of cybersecurity, navigating the complexities of risk management demands more than just vigilance and robust security measures; it requires a compass that can guide strategic decision-making and operational effectiveness. This compass is found in the form of Key Performance Indicators (KPIs), the quantifiable metrics that provide critical insights into the efficacy of an organization's cybersecurity posture. KPIs in cybersecurity risk management transform abstract risks into tangible data, allowing organizations to assess their security stance, measure the impact of their defenses, and make informed decisions that align with their strategic objectives and risk appetite.

The strategic importance of KPIs transcends mere measurement; they are pivotal in shaping an organization's approach to cybersecurity, from identifying vulnerabilities and threats to optimizing response strategies and resource allocation. By translating complex cybersecurity data into actionable intelligence, KPIs empower organizations to steer their security efforts more precisely, ensuring alignment with business goals and regulatory requirements.

As we delve into this chapter, we aim to illuminate the intricacies of selecting, implementing, and leveraging KPIs in cybersecurity risk management. Through a comprehensive exploration, we will uncover how these indicators can significantly enhance an organization's ability to navigate the cybersecurity landscape with confidence and strategic foresight.

DOI: 10.1201/9781003562184-19

UNDERSTANDING THE ROLE OF KPIs IN CYBERSECURITY RISK MANAGEMENT

Defining KPIs

KPIs in the context of cybersecurity risk management are specific, quantifiable metrics designed to evaluate the effectiveness and efficiency of an organization's cybersecurity initiatives. These metrics are carefully selected to reflect the organization's security objectives, providing a data-driven foundation for assessing performance against predefined targets. KPIs serve as a bridge between abstract security concepts and measurable outcomes, enabling organizations to gauge their resilience against cyber threats in real time.

Importance of KPIs

The role of KPIs in cybersecurity extends beyond mere measurement; they are instrumental in informing strategic decision-making and operational planning. By offering insights into the performance of cybersecurity practices, KPIs help organizations identify areas of strength and pinpoint vulnerabilities needing attention. This actionable intelligence is crucial for prioritizing security efforts, allocating resources effectively, and adjusting strategies in response to evolving threats. Furthermore, KPIs facilitate clear communication about cybersecurity performance and risks among stakeholders, enhancing transparency and fostering a culture of continuous improvement.

In the next section, we'll explore the critical considerations and criteria for selecting the right KPIs for cybersecurity, ensuring they align with organizational goals and effectively measure cybersecurity performance.

SELECTING THE RIGHT KPIs FOR CYBERSECURITY

The selection of appropriate KPIs is a pivotal step in cybersecurity risk management, demanding a strategic approach to ensure that the metrics not only align with organizational goals but also provide actionable insights. The process involves a careful balance between broad security objectives and the specificity needed for effective measurement and action.

Criteria for selection

Selecting the right KPIs requires consideration of several key factors:

- **Relevance to Business Objectives:** KPIs should be directly aligned with the organization's overarching business goals and security objectives. This alignment ensures that cybersecurity efforts are focused on protecting critical assets and functions that matter most to the organization.
- **Measurability:** Effective KPIs must be quantifiable, offering a clear metric that can be tracked over time. This quantifiability allows for objective assessment and comparison, facilitating informed decision-making.
- **Ability to Drive Action:** The chosen KPIs should provide insights that prompt specific actions or interventions. Metrics that highlight areas of risk or inefficiency can help direct resources and efforts where they are needed most, enhancing the overall security posture.

Examples of effective KPIs

To illustrate how KPIs can be effectively employed in cybersecurity risk management, let's consider a few examples:

- **Number of Detected Incidents:** This KPI measures the frequency of detected security incidents within a given timeframe, offering insights into the effectiveness of an organization's detection capabilities.
- **Average Time to Detect and Respond to Threats (Detection and Response Time):** A critical measure of how swiftly an organization can identify and mitigate threats, this KPI reflects the efficiency of the incident response process.
- **Percentage of Systems Compliant with Security Policies:** This metric assesses the extent to which IT systems adhere to established security policies, indicating the level of internal compliance and potential vulnerabilities.
- **Security Training Completion Rate:** By measuring the percentage of employees who have completed mandatory security training, this KPI evaluates the organization's commitment to fostering a culture of security awareness.

By carefully selecting KPIs based on these criteria, organizations can ensure that their cybersecurity metrics provide meaningful insights that contribute to strategic decision-making and operational improvement. The next section will delve into the implementation and monitoring of these KPIs to ensure they effectively contribute to cybersecurity risk management.

IMPLEMENTING AND MONITORING KPIs

Once the right KPIs for cybersecurity have been selected, the next critical steps are their effective implementation and continuous monitoring. These stages are vital to ensure that the KPIs serve their intended purpose of guiding cybersecurity strategies and operations.

Integration into risk management processes

For KPIs to have a meaningful impact, they must be seamlessly integrated into the organization's existing risk management frameworks and processes. This integration ensures that KPI tracking becomes an integral part of the cybersecurity risk management lifecycle, from assessment through to mitigation and review.

- **Embedding KPIs in Daily Operations:** To facilitate this, cybersecurity KPIs should be made part of the daily operational monitoring, ensuring that they are consistently measured and reviewed by the relevant teams.
- **Alignment with Risk Assessment:** KPIs should be directly tied to the outcomes of risk assessments, allowing for the dynamic adjustment of security measures in response to changing risk landscapes.
- **Stakeholder Engagement:** Engaging key stakeholders in the process of KPI monitoring and review is crucial. Regular updates and reports on KPI performance should be shared with stakeholders to ensure transparency and foster a culture of continuous improvement.

Tools and technologies

Leveraging the right tools and technologies is essential for the effective tracking and analysis of KPIs. From Security Information and Event Management (SIEM) systems to advanced analytics platforms, the choice of technology can significantly impact the organization's ability to monitor KPIs efficiently.

- **SIEM Systems:** These systems can aggregate and analyze log data from various sources, providing real-time monitoring and alerts that can be mapped to specific KPIs.
- **Advanced Analytics and Machine Learning:** For more sophisticated KPIs, especially those involving predictive analysis, machine learning (ML) models can be trained to identify patterns and anomalies that might indicate a deviation from expected KPI performance.

Continuous monitoring and adjustment

The cybersecurity landscape is dynamic, with new threats and vulnerabilities emerging continually. Consequently, KPIs must be regularly reviewed and adjusted to remain relevant and effective.

- **Regular Review Cycles:** Establishing a schedule for regular KPI reviews allows the organization to adjust its cybersecurity strategies based on the latest data and insights.
- **Adaptability:** The organization must be prepared to modify or replace KPIs as the cybersecurity environment evolves. This adaptability ensures that KPIs continue to provide relevant and actionable insights.

Implementing and continuously monitoring KPIs are foundational to ensuring that these metrics effectively guide and improve cybersecurity risk management efforts. By integrating KPIs into risk management processes, utilizing the appropriate tools for monitoring, and maintaining flexibility to adjust KPIs as needed, organizations can significantly enhance their cybersecurity posture.

In the following section, we'll explore how organizations can leverage KPIs to not only measure performance but also drive significant improvements in their cybersecurity posture, highlighting the strategic decisions informed by KPI insights.

LEVERAGING KPIs TO ENHANCE CYBERSECURITY POSTURE

Once effectively integrated and monitored, KPIs can be a powerful tool in driving strategic decisions and operational improvements in cybersecurity. This section explores how organizations can leverage these metrics to strengthen their cybersecurity posture and foster a proactive security culture.

Driving strategic decisions

KPIs serve as a critical input for strategic decision-making in cybersecurity, enabling organizations to allocate resources more effectively, prioritize cybersecurity initiatives, and justify investments in security technologies.

- **Resource Allocation:** By analyzing KPIs, organizations can identify areas of cybersecurity that require additional resources or areas where resources can be optimized for better use. For instance, a high number of security incidents in a particular department might indicate the need for additional training or more robust security measures.
- **Prioritizing Security Initiatives:** KPIs help in determining which cybersecurity projects should be prioritized based on their potential impact on improving security metrics. Projects that directly contribute to improving lagging KPIs can be moved up in priority.
- **Justifying Investments:** With concrete data from KPIs, cybersecurity teams can better justify the need for investments in specific security tools or technologies to stakeholders by demonstrating how these investments could improve key metrics.

Improving incident response

KPIs related to incident detection and response time are invaluable in assessing and improving an organization's ability to manage and mitigate cyber threats effectively.

- **Reducing Detection and Response Times:** Continuous monitoring of KPIs like average time to detect (TTD) and time to respond (TTR) to incidents can highlight inefficiencies in the incident response process. Organizations can then implement targeted improvements to reduce these times, enhancing their resilience against attacks.
- **Benchmarking Performance:** Organizations can benchmark their incident response times against industry standards or past performance to gauge their progress and identify areas for improvement.

Fostering a culture of security awareness

KPIs can also play a significant role in raising awareness about cybersecurity risks and promoting a culture of security within the organization.

- **Engagement and Training:** Security training completion rates and phishing simulation success rates are examples of KPIs that can drive engagement and participation in cybersecurity training programs. Improvements in these KPIs can indicate a growing culture of security awareness among employees.
- **Stakeholder Communication:** Regularly communicating KPIs to stakeholders across the organization helps in reinforcing the importance of cybersecurity and encourages a shared responsibility for maintaining security.

By strategically leveraging KPIs, organizations can not only measure but also significantly improve their cybersecurity posture. KPIs offer a roadmap for making informed strategic decisions, enhancing incident response capabilities, and building a strong culture of security awareness.

In the next section, we'll present real-world case studies that illustrate how organizations have successfully leveraged KPIs in their cybersecurity initiatives, showcasing the tangible benefits and lessons learned from these experiences.

CASE STUDIES: KPIs IN ACTION

Examining real-world applications of KPIs in cybersecurity provides valuable insights into their practical benefits and challenges. This section presents a selection of case studies that highlight how organizations have successfully leveraged KPIs to drive significant improvements in their cybersecurity risk management practices.

Success story: Banking Sector

Background

A leading global bank faced increasing cybersecurity threats targeting its vast network of retail banking operations. Despite a substantial cybersecurity budget, the bank struggled to demonstrate improvements in its security posture.

Implementation of KPIs

The bank adopted a set of strategic KPIs focused on key areas such as incident detection time, response time, and the rate of false positives in threat detection.

Outcomes

- **Enhanced Detection and Response:** By focusing on reducing detection and response times, the bank was able to significantly improve its ability to mitigate threats before they could impact customers or operations.
- **Optimized Resource Allocation:** The reduction in false positive rates led to more efficient use of security resources, allowing the cybersecurity team to focus on genuine threats.
- **Improved Stakeholder Confidence:** Regular reporting on these KPIs to the board and stakeholders demonstrated the value of the cybersecurity investments, leading to sustained support for security initiatives.

Learning from challenges: Healthcare provider

Background

A large healthcare provider implemented a comprehensive suite of cybersecurity measures but found it challenging to assess the effectiveness of these measures in protecting patient data.

Implementation of KPIs

The organization introduced KPIs related to data breach incidents, system compliance rates, and employee cybersecurity training completion.

Challenges

- **Data Breach Incident Rates:** Despite initial improvements, the organization noticed a plateau in progress, with incident rates no longer declining.
- **Training Completion Rates:** While training completion rates were high, this did not translate into a proportional decrease in phishing attack success rates among staff.

Lessons learned

- **Beyond Numbers:** The organization learned the importance of not just tracking KPIs but also analyzing the underlying causes behind the numbers. This led to a revamp of the training programs to focus more on practical, scenario-based training.
- **Continuous Improvement:** The experience underscored the need for continuous evaluation and adaptation of KPIs to reflect changing cybersecurity landscapes and organizational goals.

These case studies underscore the critical role KPIs play in not only measuring but actively improving an organization's cybersecurity posture. Through careful selection, implementation, and ongoing evaluation of KPIs, organizations can gain valuable insights into their security practices, driving strategic improvements and fostering a culture of security awareness and continuous improvement.

Next, we'll explore future directions for KPIs in cybersecurity, considering emerging trends and technologies that could shape the evolution of risk management strategies.

FUTURE DIRECTIONS: EVOLVING KPIs FOR CYBERSECURITY

As the cybersecurity landscape continues to evolve, so too must the KPIs that organizations rely on to measure and manage their security posture. This section delves into the anticipated changes in the cyber threat environment, emerging technologies, and how these factors might influence the development and application of KPIs in cybersecurity risk management.

Anticipating changes in the cyber threat landscape

The cyber threat landscape is dynamic, with new threats, vulnerabilities, and attack vectors emerging constantly. Organizations must stay ahead of these changes by:

- **Adapting KPIs to New Threats:** As cyber threats evolve, KPIs must be revisited and revised to ensure they remain relevant and focused on current and emerging risks. For example, the rise of ransomware attacks might necessitate KPIs specifically measuring an organization's preparedness and response to such threats.
- **Predictive Analytics:** Leveraging advanced analytics and ML to develop predictive KPIs could become crucial. These KPIs could forecast potential security incidents based on current trends, enabling proactive rather than reactive cybersecurity strategies.

Leveraging emerging technologies

Technological advancements offer new tools and approaches for cybersecurity but also introduce new complexities. Organizations should consider:

- **Integration of AI and Machine Learning:** As artificial intelligence (AI) and ML technologies become more sophisticated, they can play a significant role in enhancing KPIs related to threat detection, incident response, and predictive analytics. However, they also necessitate KPIs that measure the effectiveness and accuracy of these AI-driven systems.
- **Quantum Computing:** With the advent of quantum computing, current encryption standards and security measures may become obsolete. Future KPIs will need to assess an organization's readiness for quantum-resistant cryptography and other quantum-era security challenges.

Adaptive strategies for KPI evolution

To ensure KPIs remain effective tools for cybersecurity risk management, organizations must adopt adaptive strategies that allow for the continuous evolution of KPIs:

- **Regular KPI Review Processes:** Establishing regular review cycles for evaluating and updating KPIs ensures they stay aligned with the organization's changing risk profile and cybersecurity landscape.

- **Stakeholder Engagement:** Continuous dialogue with stakeholders across the organization ensures KPIs reflect broader business objectives and challenges, fostering a unified approach to cybersecurity.
- **Flexibility and Scalability:** Designing KPIs to be flexible and scalable allows for adjustments as new technologies are adopted and as the organization grows or changes.

CONCLUSION

KPIs are indispensable tools in the arsenal of cybersecurity risk management, offering clarity and direction in the complex and ever-evolving cyber threat landscape. As we've explored throughout this chapter, the selection, implementation, and leveraging of KPIs require strategic thought, continuous adaptation, and a deep understanding of both the organization's goals and the cybersecurity challenges it faces.

Looking ahead, the future of cybersecurity KPIs will be shaped by advances in technology, changes in the threat landscape, and the growing sophistication of cyber adversaries. Organizations that remain agile, continuously evaluating and adapting their KPIs, will be better positioned to navigate these changes, safeguard their digital assets, and maintain resilience against cyber threats.

In navigating the cybersecurity challenges of tomorrow, organizations must not only embrace the strategic use of KPIs but also foster a culture of innovation, collaboration, and continuous improvement. By doing so, they can turn the complexity of the cyber world into a navigable path toward a more secure and confident digital future.

CYBER CHUCKLES: PRECISION NAVIGATION THROUGH THE CYBERSECURITY JUNGLE

Welcome to the wilds of the cybersecurity jungle, where the paths are as tangled as old headphone cords and the threats lurk behind every tree. But fear not, intrepid explorer, for we've ditched the unreliable compass of yesteryear and upgraded to the GPS of cybersecurity: KPIs.

The map to treasure

Imagine KPIs as the treasure map marking the X on your cybersecurity efforts. Each KPI is a step closer to the treasure chest of data protection and risk management. But beware, not all maps lead to gold; some just circle back to the start. Choosing the right KPIs is like selecting the perfect trail – some offer scenic views into the depths of your security protocols, while others are steep climbs that test your resilience.

Packing your backpack

Venturing into the cybersecurity wilderness without the right gear? Unthinkable! KPIs are the multi-tool, water filter, and compass all rolled into one. They guide you through the dense underbrush of data breaches and across the raging rivers of ransomware. With KPIs, you're not just surviving; you're setting up camp and thriving in the great digital outdoors.

Real-world explorers: Tales of triumph and tribulation

Let's sit by the campfire and share tales of those who've navigated these trails before us. First, the story of a bank that scaled the peak of Mount SecureNet, their KPIs lighting the way like trusty torches. They measured, monitored, and mastered their cybersecurity landscape, turning potential pitfalls into paths of victory.

Then, the cautionary tale of a healthcare provider who, without a clear KPI compass, wandered into the thicket of complacency. They learned the hard way that even the most fortified castles need a map to navigate the ever-changing siege tactics of cyber marauders.

The KPI compass: Pointing north to Cybersecurity Nirvana

Our journey through the cybersecurity wilderness teaches us the power of the right tools. KPIs are more than mere numbers; they're the stars guiding our ships through the stormy seas of the digital age. By measuring our progress, illuminating our successes, and revealing the dragons lurking in the depths, KPIs steer us toward the fabled land of Cybersecurity Nirvana.

In conclusion: The path well chosen

As we pack our bags and douse the campfire, let's remember the lessons of the trails. With KPIs in hand, the cybersecurity jungle becomes less daunting, the paths clearer, and the treasure within our reach. So, here's to choosing our KPIs wisely, navigating with precision, and embarking on cybersecurity adventures with confidence. May our digital expeditions be fruitful, our data protected, and our risks managed. Onward to the horizon, where the cybersecurity sun never sets!

And so, armed with our KPI compass, we step back into the wild, ready to chart a course through the cybersecurity jungle, from the peaks of prevention to the valleys of risk management, all the while keeping our eyes on the treasure of security and resilience.

Chapter 15

Adapting KANBAN for cybersecurity excellence

Spoiler Alert!

- This chapter explores how the Kanban method, traditionally used for project management, can be a powerful tool for cybersecurity teams.
- Imagine a giant whiteboard with sticky notes representing your cybersecurity tasks – that's Kanban in action!
- This chapter dives into the core principles of Kanban and how to implement them in your cybersecurity team (think of it as setting up your Kanban board for maximum efficiency).
- Learn how to visualize workflows, prioritize tasks, and adapt to changing threats (because the cybersecurity landscape is anything but static).
- This chapter also explores the challenges of implementing Kanban in cybersecurity (it's not always smooth sailing!).
- Get ready for some real-world examples: a financial services firm using Kanban to great success and a healthcare provider facing some Kanban roadblocks (because learning happens from both victories and defeats).

THIS WAY PLEASE

In the fast-paced world of digital security, where threats evolve at an unprecedented rate, cybersecurity teams are constantly seeking methodologies to enhance their agility, effectiveness, and resilience. Among the various management approaches that have transformed industries, Kanban stands out for its simplicity, adaptability, and focus on continuous improvement. Originally developed to streamline manufacturing processes, Kanban has found relevance in a wide range of fields, including software development and, more recently, cybersecurity.

This chapter explores the integration of Kanban principles into the fabric of cybersecurity management. It posits that adopting Kanban can offer new perspectives on managing the intricate tasks of safeguarding sensitive data and mitigating digital risks. Through visual management, the pull system, and an emphasis on continuous improvement, Kanban holds the potential to revolutionize how cybersecurity teams operate, adapt, and collaborate.

By delving into the core principles of Kanban and illustrating their applicability to the realm of information security, this chapter aims to provide a roadmap for cybersecurity teams eager to enhance their strategies and workflows. The goal is to showcase how embracing Kanban can lead to a more responsive, efficient, and robust cybersecurity posture, paving the way for a secure digital future.

DOI: 10.1201/9781003562184-20

THE CONVERGENCE OF KANBAN AND CYBERSECURITY

The fusion of Kanban principles with cybersecurity practices represents a promising frontier for enhancing digital defenses. This section breaks down how the foundational elements of Kanban can be adapted to the cybersecurity context, providing a framework for more effective management of security tasks and initiatives.

Kanban's core principles

Kanban, with its emphasis on visual representation, continuous improvement, flexibility, and the pull principle, provides a robust framework for managing complex processes. In the context of cybersecurity:

- **Visual Representation:** Implementing a Kanban board to map out cybersecurity tasks – from identifying vulnerabilities to managing ongoing security projects – can clarify priorities and progress. This visual tool helps in identifying bottlenecks, enabling timely interventions.
- **Continuous Improvement:** Cyber threats never stand still, nor should cybersecurity practices. Kanban encourages a culture where teams regularly evaluate and refine their workflows, ensuring that security measures evolve to counter new and emerging threats effectively.
- **Pull Principle:** Adopting a pull-based approach allows cybersecurity tasks to be addressed based on priority and team capacity, ensuring critical issues are tackled promptly without overloading resources.
- **Flexibility and Adaptability:** The dynamic nature of cyber threats necessitates a flexible and adaptable management approach. Kanban's inherent flexibility allows cybersecurity teams to rapidly adjust their focus and resources in response to emerging threats or changing priorities.

Visualizing cybersecurity workflows

The adoption of Kanban starts with translating cybersecurity workflows into a visual format. A Kanban board, segmented into various stages of task completion (such as "To Do," "In Progress," and "Done"), provides a clear overview of all ongoing and pending security tasks. This visualization aids in prioritizing actions, tracking progress, and quickly identifying areas requiring immediate attention or adjustment.

In the following sections, we'll delve deeper into the implementation of Kanban in cybersecurity teams, addressing the practical steps involved, the benefits of this approach, and the challenges teams may encounter.

IMPLEMENTING KANBAN IN CYBERSECURITY TEAMS

Implementing Kanban within cybersecurity teams involves a thoughtful approach to adapting its principles to the unique challenges and workflows of information security management. This section outlines the steps for creating a visual workflow, applying the pull principle, and ensuring flexibility and adaptability in cybersecurity operations.

Creating a visual workflow

The first step in adopting Kanban for cybersecurity is to establish a visual representation of the team's workflow. This visualization, typically in the form of a Kanban board, is crucial for managing the array of tasks that cybersecurity teams face daily.

- **Designing the Kanban Board:** Start by mapping out the key stages of your cybersecurity workflow on a Kanban board. This might include columns such as "Backlog," "Analysis," "In Progress," "Review," and "Completed." The board can be physical, with sticky notes or cards, or digital, using one of the many available Kanban software tools.
- **Populating the Board:** Once the board is set up, populate it with tasks, represented by cards or notes. Each task should include a clear description, the assigned team member (if applicable), and any relevant deadlines or priorities.
- **Customizing to Fit Your Needs:** The Kanban board should be customized to reflect the specific processes and terminologies of your cybersecurity team. Over time, you may find that certain stages need to be added, removed, or modified to better suit your team's workflow.

Applying the pull principle

In Kanban, work is "pulled" through the system as capacity allows, rather than being "pushed" based on a predetermined schedule. This principle is particularly beneficial for cybersecurity teams, where the workload can be unpredictable and urgent tasks often arise.

- **Prioritizing Tasks:** Use the Kanban board to prioritize tasks based on their urgency and importance. High-priority issues, such as active security breaches or critical vulnerabilities, should be moved to the front of the queue.
- **Managing Work in Progress (WIP):** To prevent overload and ensure that critical tasks receive the attention they need, limit the number of tasks in the "In Progress" stage. This WIP limit helps maintain focus and improves the quality of work.
- **Pulling Based on Capacity:** Team members pull new tasks from the backlog or the next stage of the workflow only when they have the capacity to do so. This approach ensures that tasks are addressed promptly and efficiently without overwhelming the team.

Emphasizing flexibility and adaptability

The dynamic nature of cybersecurity threats requires an equally dynamic response. Kanban's inherent flexibility allows teams to adapt their workflows and priorities as the threat landscape evolves.

- **Adjusting in Real Time:** The visual nature of the Kanban board makes it easy to adjust priorities and workflows in real time, in response to emerging threats or changes in the organization's security posture.
- **Iterative Process:** Kanban encourages an iterative approach to process improvement. Regularly review the workflow and task management practices for opportunities to increase efficiency and effectiveness.
- **Incorporating Feedback and Learning:** Foster an environment where team members can provide feedback on the workflow and suggest improvements. This collective input is invaluable for refining the Kanban system to better meet the needs of the cybersecurity team.

By carefully implementing these steps, cybersecurity teams can leverage Kanban to enhance their operational efficiency, improve task prioritization, and foster a culture of continuous improvement. In the next section, we'll explore the tangible benefits and challenges of applying Kanban principles to cybersecurity management, drawing on real-world examples to illustrate these concepts in action.

CHALLENGES AND CONSIDERATIONS

While Kanban offers a flexible and efficient framework for managing tasks and workflows, its application in the cybersecurity domain presents unique challenges and considerations. This section addresses potential obstacles and provides guidance on customizing Kanban to fit the intricacies of cybersecurity operations effectively.

Customization to cybersecurity needs

Kanban's principles, while versatile, may not directly align with the specific demands of cybersecurity work without thoughtful adaptation.

- **Tailoring Kanban Boards:** Cybersecurity tasks vary in complexity, urgency, and impact. The design of Kanban boards should reflect these nuances, possibly requiring multiple boards or specialized columns to categorize tasks effectively (e.g., immediate threats, ongoing vulnerabilities, compliance tasks).
- **Sensitive Information Handling:** Given the sensitive nature of cybersecurity work, precautions should be taken when visualizing tasks. Consider the level of detail displayed on cards and choose secure, compliant tools for digital Kanban boards.

Balancing agility with rigor

Cybersecurity operations must be agile to respond to threats swiftly but also rigorous to ensure thorough protection and compliance with standards.

- **Agility vs. Thoroughness:** Kanban promotes responsiveness and flexibility, which are crucial for adapting to new threats. However, it's essential to balance this with the need for comprehensive risk assessments and methodical incident response plans.
- **Maintaining Compliance:** Implementing Kanban should not compromise an organization's adherence to cybersecurity frameworks and regulations. Ensure that the workflow allows for necessary reviews and documentation to maintain compliance.

Integration with existing processes

Successfully integrating Kanban into existing cybersecurity processes requires careful planning and communication.

- **Alignment with Security Frameworks:** Align Kanban practices with the organization's established cybersecurity frameworks and protocols. This alignment ensures that Kanban enhances rather than disrupts effective security measures.
- **Stakeholder Buy-in:** Gaining the support of stakeholders, from IT staff to executive management, is crucial for the successful adoption of Kanban. Communicate the

benefits clearly and demonstrate how Kanban can improve efficiency and outcomes in cybersecurity operations.

Overcoming resistance to change

Change can be challenging, and introducing a new system like Kanban may meet resistance within the team or organization.

- **Training and Education:** Provide comprehensive training on Kanban principles and practices, emphasizing their benefits for cybersecurity work. Highlight successful case studies and potential for improved team collaboration and efficiency.
- **Phased Implementation:** Consider a phased approach to implementing Kanban, starting with pilot projects or specific areas of the cybersecurity operation. This allows the team to adjust to the new system gradually and provides opportunities to refine the approach based on early experiences.

By acknowledging these challenges and considerations, cybersecurity teams can better prepare for the successful integration of Kanban principles into their operations. The key lies in customization, communication, and a commitment to continuous improvement, ensuring that Kanban not only fits the unique landscape of cybersecurity but also enhances the team's ability to protect against digital threats effectively.

In the next section, we will explore case studies that highlight the successful application of Kanban principles in cybersecurity settings, providing insights into the practical benefits and lessons learned from these experiences.

CASE STUDIES: KANBAN IN ACTION FOR CYBERSECURITY

Real-world applications of Kanban in cybersecurity provide valuable insights into the methodology's practical benefits and implementation challenges. These case studies showcase organizations that have successfully integrated Kanban principles into their cybersecurity operations, highlighting the strategies employed, the obstacles overcome, and the tangible improvements realized.

Success story: Financial services firm

Background

A mid-sized financial services firm faced challenges in managing a growing volume of cybersecurity tasks, including threat monitoring, incident response, and compliance activities. The cybersecurity team often found itself reacting to issues rather than proactively managing risks.

Implementation of Kanban

- **Visual Workflow Creation:** The team created a Kanban board with stages tailored to their specific processes, including threat analysis, mitigation actions, and post-incident reviews.
- **Prioritization and Pull Principle:** High-priority tasks, such as active threats and regulatory compliance deadlines, were clearly highlighted and addressed based on their urgency and the team's capacity, ensuring a focused and efficient response.

- **Continuous Improvement:** Regular retrospective meetings were instituted to discuss completed tasks, identify bottlenecks, and explore opportunities for workflow optimization.

Outcomes

- **Enhanced Response Times:** The visual nature of the Kanban board and the prioritization of tasks led to a significant reduction in response times to security incidents.
- **Improved Team Collaboration:** The board facilitated better communication and collaboration within the team and with other departments, leading to a more cohesive approach to cybersecurity.
- **Increased Operational Efficiency:** Continuous improvement efforts helped streamline processes, reducing the time and resources required to manage cybersecurity tasks.

Learning from challenges: Healthcare provider

Background

A large healthcare provider sought to improve its information security practices in the face of increasing cyber threats and stringent regulatory requirements. Despite having a skilled cybersecurity team, managing and prioritizing the vast array of tasks proved daunting.

Implementation of Kanban

- **Adapting Kanban for Sensitive Operations:** Given the sensitivity of patient data, the team carefully designed their Kanban board to ensure confidentiality, using coded task descriptions and secure digital tools.
- **Flexibility in Task Management:** The team applied Kanban's flexibility principle to adjust their focus rapidly in response to emerging threats or changes in compliance requirements.
- **Stakeholder Engagement:** To foster a broader culture of cybersecurity awareness, the Kanban system was integrated with regular communication and training sessions for non-security staff.

Challenges

- **Cultural Resistance:** Initial resistance to the new system was encountered, particularly from team members accustomed to traditional, hierarchical task assignments.
- **Complex Regulatory Environment:** Navigating the complex landscape of healthcare regulations while maintaining the agility promoted by Kanban required ongoing adjustments to the team's workflow and priorities.

Lessons learned

- **Engagement and Training Are Key:** Addressing resistance through comprehensive training and demonstrating Kanban's benefits were crucial for gaining team buy-in.
- **Customization Is Crucial:** The experience underscored the importance of customizing Kanban to fit the unique demands and regulatory context of the healthcare sector.

These case studies illustrate the adaptability of Kanban principles across different sectors and cybersecurity challenges. By embracing Kanban's core tenets – visual management, continuous improvement, and flexibility –organizations can enhance their cybersecurity posture, streamline operations, and foster a proactive security culture.

In the concluding section, we will reflect on the broader implications of adopting Kanban for cybersecurity management and explore future directions for this promising approach.

MANAGEMENT METHODS OVERVIEW

Before finishing this chapter, let's briefly explore how it stacks up against other prominent management methodologies based on creativity, dynamism, size, complexity, and culture (Table 15.1).

1. Lean Management

- **Creativity:** Low to Medium
- **Dynamism:** Low to Medium
- **Size:** Small to Medium
- **Complexity:** Simple to Medium
- **Culture:** Efficiency-oriented, structured

Lean Management focuses on minimizing waste and optimizing processes, making it ideal for environments that prioritize efficiency over flexibility.

2. Kanban

- **Creativity:** Medium to High
- **Dynamism:** High
- **Size:** Small to Medium
- **Complexity:** Medium
- **Culture:** Flexible, visually oriented, continuous improvement

Kanban's strength lies in its adaptability and visual workflow management, promoting continuous improvement and flexibility, which are crucial in dynamic environments like cybersecurity.

3. Scrum

- Creativity: High
- Dynamism: High

Table 15.1 Detailed overview table

Method	*Creativity*	*Dynamism*	*Size*	*Complexity*	*Culture*
Lean Management	Low-Medium	Low-Medium	Small-Medium	Simple-Medium	Efficiency-oriented, structured
Kanban	Medium-High	High	Small-Medium	Medium	Flexible, visually oriented
Scrum	High	High	Small-Medium	Medium-High	Team-oriented, iterative
Six Sigma	Low-Medium	Low	Medium-Large	Medium-High	Data-driven, quality-oriented
PRINCE2	Low-Medium	Low-Medium	Medium-Large	High	Process-driven, formal
Agile (General)	High	High	Small-Medium	Medium-High	Flexible, adaptable
TOC	Medium	Medium-High	Medium-Large	Medium-High	Bottleneck-oriented, optimizing

- **Size:** Small to Medium
- **Complexity:** Medium to High
- **Culture:** Team-oriented, iterative, self-organized

Scrum is highly iterative and collaborative, making it suitable for projects that require high creativity and rapid adaptation to change.

4. Six Sigma

- **Creativity:** Low to Medium
- **Dynamism:** Low
- **Size:** Medium to Large
- **Complexity:** Medium to High
- **Culture:** Data-driven, quality-oriented, structured

Six Sigma is a data-driven approach focused on quality improvement and defect reduction, best suited for large organizations with complex processes.

5. PRINCE2

- **Creativity:** Low to Medium
- **Dynamism:** Low to Medium
- **Size:** Medium to Large
- **Complexity:** High
- **Culture:** Process-driven, formal, clearly defined roles

PRINCE2 is a process-driven methodology with a strong emphasis on clear roles and responsibilities, ideal for large projects with high complexity.

6. Agile (General)

- **Creativity:** High
- **Dynamism:** High
- **Size:** Small to Medium
- **Complexity:** Medium to High
- **Culture:** Flexible, adaptable, collaborative

Agile methodologies emphasize flexibility, adaptability, and collaboration, making them suitable for dynamic and innovative environments.

7. Theory of Constraints (TOC)

- **Creativity:** Medium
- **Dynamism:** Medium to High
- **Size:** Medium to Large
- **Complexity:** Medium to High
- **Culture:** Bottleneck-oriented, optimizing

TOC focuses on identifying and addressing bottlenecks to optimize the overall system performance, which is valuable for improving efficiency in complex environments (Table 15.1).

CONCLUSION

The journey through the adaptation of Kanban principles to cybersecurity management reveals a promising pathway toward enhanced operational efficiency, improved responsiveness to threats, and a strengthened security posture. As the case studies illustrate, organizations across various sectors have successfully leveraged Kanban to navigate the complex and dynamic landscape of cybersecurity, achieving tangible benefits in the process.

The integration of Kanban into cybersecurity practices represents more than just an operational shift; it signifies a cultural transformation toward agility, transparency, and continuous improvement. By making cybersecurity tasks visible, prioritizing based on real-time demands, and fostering an environment of collaboration and adaptability, teams are better equipped to respond to the ever-evolving nature of cyber threats.

Key takeaways

- **Visual Management Empowers Teams:** The visual representation of tasks and workflows is a powerful tool for enhancing situational awareness, identifying bottlenecks, and facilitating efficient resource allocation.
- **Continuous Improvement Drives Progress:** The principle of kaizen, or continuous improvement, is crucial in an area as dynamic as cybersecurity. Regular reviews and adaptations of security processes ensure that defenses remain robust and effective.
- **Flexibility and Responsiveness are Essential:** In the face of rapidly emerging threats, the ability to adapt and respond quickly is paramount. Kanban's emphasis on flexibility and the pull principle supports a proactive and agile security posture.
- **Collaboration Enhances Security Culture:** The transparency and collaborative nature of Kanban fosters a culture of shared responsibility for cybersecurity, engaging all team members in the collective effort to safeguard digital assets.

FUTURE DIRECTIONS

As cybersecurity challenges continue to grow in complexity and scale, the adoption of agile and adaptable management methodologies like Kanban will likely become increasingly critical. Future advancements in digital technologies and the evolving threat landscape will necessitate ongoing refinement and customization of Kanban practices to meet the specific needs of cybersecurity teams.

- **Technology Integration:** Leveraging emerging technologies, such as AI and machine learning, within the Kanban framework can enhance threat detection and response capabilities.
- **Broader Organizational Adoption:** Expanding the principles of Kanban beyond cybersecurity teams to encompass all aspects of organizational security can create a more cohesive and integrated approach to risk management.
- **Enhanced Metrics and KPIs:** Developing sophisticated metrics and KPIs that align with Kanban practices will enable more nuanced assessment and management of cybersecurity risks.

In conclusion, while Kanban originated in the manufacturing sector, its principles offer valuable insights and methodologies for cybersecurity management. By embracing Kanban, organizations can cultivate a culture of continuous improvement, adaptability, and collaboration, setting the stage for a more secure digital future. The journey of integrating Kanban into cybersecurity is ongoing, with each iteration and adaptation bringing us closer to mastering the art of agile security management in an increasingly complex cyber landscape.

CYBER CHUCKLES: CYBERSECURITY'S NEW FRIEND – KANBAN

Welcome aboard the Kanban Express, where cybersecurity tasks get a first-class ticket to Efficiencyville! Imagine transforming the chaotic jungle of cyber threats into a neatly organized board of sticky notes. That's right, we're talking about applying the ancient art of Kanban to the whirlwind world of cybersecurity.

Kanban: Not just for post-its anymore

Gone are the days when Kanban was just for tracking your grocery list. Now, it's cybersecurity's turn to embrace this visually stunning method of madness management. Picture this: a giant board festooned with colorful sticky notes, each representing a task to keep digital doom at bay.

The great wall of tasks

Setting up your cybersecurity Kanban board is like planning the layout of a digital fortress. Each column is a bastion against chaos, from "New Threats on the Horizon" to "Vanquished Vulnerabilities." And those sticky notes? Your army of tasks, marching from one column to the next, battling cyber beasts along the way.

Choosing your trail

Picking the right Kanban columns is akin to choosing the perfect hiking trail. Some paths are scenic, winding through "Routine Checks" and "Ongoing Monitoring." Others are steep, leading to "Critical Vulnerabilities" and "Emergency Patches." Choose wisely, and you might just enjoy the journey.

The gear for the trek

No adventurer sets off without the right gear, and the same goes for cybersecurity Kanban warriors. This chapter isn't just about setting up your board; it's about arming yourself with the strategies to use it effectively. Think of it as packing your backpack with the essentials for cyber-survival.

Tales from the trenches

What's a good strategy without some war stories? We've got tales of triumph from a financial firm that turned its cyber-fortunes around with Kanban, and a cautionary tale from a healthcare provider that stumbled over Kanban's sticky notes. Spoiler alert: there's learning in both victory and defeat.

Conclusion: Kanban conquers chaos

As we wrap up this chapter, remember that Kanban in cybersecurity isn't just about keeping things neat and tidy. It's about transforming your team into a lean, mean, cyber-threat-fighting machine. With Kanban, you're not just responding to threats; you're staying two steps ahead, all with the help of a few well-placed sticky notes.

So, embrace the board, pick your path, and march your tasks toward victory. The world of cybersecurity might be complex, but with Kanban, it's a complexity that you can see, understand, and conquer. Here's to turning chaos into order, one sticky note at a time!

Chapter 16

Quantum leap?

Leverage quantum for security and efficiency

Spoiler Alert!

- Get ready for some mind-bending stuff: this chapter explores how quantum technology can revolutionize cybersecurity.
- Imagine a world with ultra-secure communication and efficient energy usage (thanks to quantum mechanics!).
- This chapter explores the challenges.

A huge shout-out to my niece, Annika Jerosch (age 13 during this writing), for her keen insights and valuable advice on this chapter. Annika, your fresh perspective and sharp observations were like a lighthouse guiding us through the quantum fog! Your contributions didn't just enhance this chapter; they supercharged it with clarity and depth. Thank you for being our guiding star in this quantum leap!

INTRODUCTION

The escalating challenges of global energy consumption demand a paradigm shift in our approach to technology. Quantum technologies, with their ability to harness the fundamental principles of quantum mechanics, hold immense promise for revolutionizing energy efficiency and paving the way for a sustainable future.

This short paper explores the potential of quantum internet and quantum blockchain to transform communication, computing, and energy management across various industries. It delves into the technical challenges that need to be addressed to fully realize the transformative power of these technologies and proposes innovative approaches from both technical and physics perspectives.

QUANTUM INTERNET: UNVEILING SECURE AND EFFICIENT COMMUNICATION

The traditional internet, while ubiquitous, suffers from inherent limitations in security and efficiency. Quantum internet, on the other hand, emerges as a groundbreaking solution, offering unparalleled security and enhanced energy efficiency for long-distance communication.

Quantum internet utilizes quantum entanglement, a unique property of quantum mechanics, to establish secure connections between distant nodes. Any attempt to intercept or manipulate these entangled quantum states would instantly destroy them, alerting the sender and receiver of the intrusion.

DOI: 10.1201/9781003562184-21

The energy efficiency of quantum internet stems from its ability to transmit information simultaneously along multiple paths, utilizing quantum superposition. This enables the establishment of secure connections with significantly fewer photons compared to classical communication methods.

TECHNICAL CHALLENGES: OVERCOMING DISTANCE BARRIERS AND ENSURING ROBUSTNESS

Despite its promise, quantum internet faces significant technical challenges. One major hurdle lies in the inherent fragility of quantum states, which degrade over distance, limiting the network's reach. To overcome this limitation, quantum repeaters are essential.

Quantum repeaters act as relay stations, amplifying and relaying quantum signals over long distances, ensuring their integrity and enabling truly global quantum communication networks. However, the development of efficient quantum repeaters remains an ongoing research challenge.

PHYSICS-BASED APPROACHES: HARNESSING QUANTUM MECHANICS

Physics research plays a crucial role in addressing the technical challenges of quantum internet. Studies are focused on developing novel materials and technologies to enhance quantum state manipulation and coherence preservation across long distances.

Quantum optics, quantum information processing, and condensed matter physics are key disciplines contributing to advancements in quantum repeaters and quantum communication protocols.

QUANTUM BLOCKCHAIN: SECURE AND ENERGY-EFFICIENT TRANSACTIONS

Blockchain technology, underpinning cryptocurrencies, revolutionized digital transactions with its tamper-proof ledger. However, current blockchain protocols are vulnerable to quantum computing attacks.

Quantum blockchain emerges as a solution, introducing quantum-resistant algorithms that can withstand the advanced capabilities of quantum computers. These algorithms safeguard the integrity of blockchain networks and protect the security of digital assets.

One area of focus is the development of quantum-resistant hash functions, essential for verifying transaction data and ensuring the integrity of the blockchain. Another important aspect is the design of quantum-resistant consensus protocols, which maintain agreement among nodes in the network.

TECHNICAL CHALLENGES: INTEGRATING QUANTUM SECURITY AND BALANCING EFFICIENCY

The integration of quantum-resistant algorithms into blockchain protocols poses technical challenges. These algorithms must be computationally efficient and compatible with existing blockchain architectures.

Furthermore, the energy efficiency of quantum blockchain needs to be carefully considered. The quantum hashing and consensus algorithms should not introduce excessive energy consumption, which would undermine the environmental benefits of blockchain technology.

PHYSICS-BASED APPROACHES: QUANTUM-SAFE CRYPTOGRAPHY

Physics research plays a critical role in developing quantum-resistant cryptography. Studies are focused on identifying and utilizing quantum phenomena that are resistant to quantum computing attacks.

One promising approach involves leveraging quantum-secure key distribution protocols to establish secure communication channels between blockchain nodes. These protocols can ensure that the exchanged keys remain unbreakable, even against quantum computers.

Another area of research is the development of quantum-resistant error correction codes, which can correct errors introduced during quantum computation and communication. These codes are crucial for ensuring the reliability of quantum algorithms used in quantum blockchain.

BRIDGING THE GAP: QUANTUM TECHNOLOGIES FOR ENERGY OPTIMIZATION IN DIVERSE INDUSTRIES

The transformative potential of quantum technologies extends far beyond communication and finance. Quantum algorithms hold immense promise for optimizing energy consumption across various industries, including transportation, manufacturing, and urban infrastructure.

In the transportation sector, quantum algorithms could revolutionize traffic flow management, optimize vehicle routing, and improve fuel efficiency. Quantum optimization techniques could streamline production processes in manufacturing, reduce waste, and enhance energy utilization in industrial facilities. Quantum simulations could support the design of more energy-efficient buildings, smart grids, and sustainable urban planning.

ENERGY CONSUMPTION OF QUANTUM COMPUTERS: A BALANCING ACT

Quantum computers hold immense promise for solving complex problems that are intractable for classical computers. However, their development and operation are not without challenges, particularly in terms of energy consumption. One of the most significant energy demands comes from the need to cool quantum systems to extremely low temperatures, typically near absolute zero (0 Kelvin or –273.15°C). This is necessary to minimize quantum decoherence, the process by which quantum properties like superposition and entanglement degrade over time.

The energy required for cooling depends on the specific quantum system and the desired level of coherence. For example, current quantum computers typically operate at temperatures around 10 microkelvin (μK), which requires cryogenic cooling systems that can reach ultra-low temperatures. These systems can consume a significant amount of energy, typically in the form of liquid helium or other cryogenic fluids.

The trade-off between energy consumption and computation efficiency is a critical consideration in the development of quantum computers. While cooling requirements place a burden on energy usage, the potential benefits of quantum computing are immense. Quantum algorithms can tackle problems that are intractable for classical computers, such as simulating complex molecular interactions, optimizing financial portfolios, and breaking modern cryptography.

The key to mitigating the energy consumption issue lies in improving the efficiency of quantum algorithms and reducing the cooling requirements. Researchers are actively pursuing both of these avenues.

IMPROVING QUANTUM ALGORITHM EFFICIENCY

One approach to reducing energy consumption is to improve the efficiency of quantum algorithms. This involves developing algorithms that can solve problems using fewer quantum resources, such as qubits and operations. Several promising advancements in quantum algorithm design have emerged in recent years, demonstrating the potential for significant efficiency gains.

REDUCING COOLING REQUIREMENTS

Another approach is to reduce the cooling requirements of quantum systems. This could involve developing new materials or techniques for cooling that are more energy efficient or seeking alternative methods of achieving the required level of coherence without the need for ultra-low temperatures.

While the energy consumption of quantum computers is a concern, the potential benefits of these technologies for solving complex problems and advancing scientific discovery are immense. The challenge is to strike a balance between energy efficiency and computational power. By continuing to invest in research on both fronts, we can harness the power of quantum computing while minimizing its environmental impact.

SUMMARY

Combining quantum mechanics with cybersecurity opens a fascinating array of possibilities beyond the well-known area of quantum encryption. Here are a few intriguing topics in this domain:

1. **Quantum Key Distribution (QKD)**: While this falls under quantum encryption, it's worth mentioning for its unique approach to secure communication, using quantum properties to create theoretically unbreakable encryption.

   The BB84 Protocol: A short glance at QKD: QKD is a revolutionary method for secure communication. Let's quickly explore the BB84 protocol:

   Protocol Details: Step-by-Step Explanation: Alice and Bob use polarized photons to exchange a secret key. Different polarization bases (rectilinear and diagonal) ensure security. Error correction and privacy amplification steps create a shared secret key.

   Implementation Challenges: The Quantum Reality Check: Photon transmission over long distances without loss or disturbance is challenging. Quantum repeaters extend the range of QKD networks, and integrating QKD with existing communication infrastructure poses additional challenges.

Real-World Examples: Where QKD Is Making a Difference: The Defense Advanced Research Projects Agency (DARPA) Quantum Network is a pioneering QKD network in the United States. Companies like ID Quantique and Toshiba are developing commercial QKD systems.

2. **Quantum Random Number Generation:** Quantum mechanics can be used to generate truly random numbers, which are a crucial component in cryptographic algorithms and securing communications.
3. **Post-Quantum Cryptography:** Developing cryptographic systems that are secure against quantum computers. This involves creating algorithms that even quantum computers find difficult to crack.
4. **Quantum Hacking and Countermeasures:** Studying potential quantum computing threats to existing cryptographic systems and devising quantum-resistant security measures.
5. **Quantum Secure Direct Communication (QSDC):** A method for direct communication of secret messages without the need for a key, using quantum entanglement and superposition principles.
6. **Quantum Internet:** Exploring the use of quantum mechanics in creating a new type of internet that would offer novel ways to achieve secure communication.
7. **Quantum Blockchain:** Applying quantum mechanics to enhance the security and efficiency of blockchain technology.

Each of these areas presents unique challenges and opportunities for research and innovation, combining the cutting edge of quantum physics with the practical needs of cybersecurity.

Focusing on points 6 and 7 could have huge benefits for energy consumption savings:

1. **Quantum Internet:** This concept involves using quantum signals instead of traditional digital signals for communication. The key feature of a quantum internet is its potential for intrinsically secure communication, based on the principles of quantum mechanics. This is because the act of measuring a quantum state (necessary for any eavesdropping) inherently alters that state, revealing any interception attempt. This technology is still in its infancy, with significant research being conducted on quantum repeaters, quantum routers, and the overall architecture needed for a quantum network.
2. **Quantum Blockchain:** The application of quantum computing to blockchain technology could significantly enhance its security and efficiency. Quantum-resistant algorithms are being explored to safeguard blockchain against potential quantum computing attacks. Additionally, the use of quantum techniques could improve the speed and scalability of blockchain transactions.

   This concept is still largely theoretical and requires advancements in both quantum computing and blockchain technology to be fully realized.

OVERSIGHT

Charting a sustainable technological future

The advent of quantum internet and quantum blockchain represents a pivotal shift in our approach to communication, computing, and energy efficiency. As research advances, overcoming current barriers, these technologies could significantly contribute to a sustainable future characterized by minimized energy consumption and fortified security.

Table 16.1 Quantum sustainability overview

Technology	*Benefits*	*Technical challenges*
Quantum internet	Ultra-secure communication over vast distances, reduced energy requirements for communication	Quantum repeaters, maintaining quantum coherence over long distances
Quantum blockchain	Quantum-resistant algorithms against quantum computers, enhanced energy efficiency in cryptocurrency mining	Integration of quantum-resistant algorithms, refining of hashing algorithms
Quantum transportation	Optimized traffic flow, reduced energy consumption in vehicles, improved fuel efficiency	Development of quantum algorithms for traffic optimization, integration of quantum sensors in vehicles
Quantum manufacturing	Streamlined production processes, reduced waste, optimized energy usage in industrial facilities	Development of quantum optimization techniques for manufacturing processes, integration of quantum control systems
Quantum urban infrastructure	Design of more energy-efficient buildings, smart grids, and sustainable urban planning	Quantum simulations for urban planning, integration of quantum sensors in buildings and infrastructure

Embracing the profound capabilities of quantum mechanics opens the door to a world where technology not only meets our present needs but does so in a way that safeguards the future of our planet and its inhabitants (Table 16.1).

CYBER CHUCKLES: QUANTUM LEAP – THE AMUSEMENT PARK OF TOMORROW'S CYBERSECURITY AND ENERGY EFFICIENCY

1. **Opening Scene**: Our enthusiastic protagonist, Professor Quantum, stands at the entrance of "Quantum Land," a futuristic theme park with attractions named after quantum concepts. A sign reads, "Welcome to Quantum Land: Where the Future is Entangled!"
2. **The Quantum Internet Tunnel**: Visitors enter a rollercoaster named "The Quantum Entanglement Express," designed to mimic ultra-secure quantum communication channels. Riders wearing VR headsets experience being quantum bits (qubits) sent securely across vast distances. A child comments, "This is cooler than WiFi!", while an elderly lady holds on tight, saying, "I just wanted to send an email!"
3. **Blockchain Bumper Cars**: In this attraction, bumper cars are equipped with "quantum shields," making them immune to collisions (symbolizing quantum blockchain's invulnerability to hacking). A sign humorously warns, "Bump at your own risk: Quantum security in place!" A frustrated hacker in the audience tries to control a car remotely, only to be outsmarted by quantum encryption.
4. **Quantum Cooling Chambers**: A funhouse maze filled with fog and mirrors, representing the challenge of keeping quantum computers cool. Visitors navigate through "cooling paths" with air gusts and end up in front of a giant, mock quantum computer covered in fake snow. A teenager jokes, "My phone overheats more than this!"
5. **The Energy-Efficient Quantum Future**: A pavilion showcasing quantum-enhanced gadgets that use minimal energy, including a quantum lightbulb so efficient it's powered by the mere presence of visitors. A family marvels at a quantum-powered home model, while a dad tries to power the lightbulb with his dance moves, exclaiming, "I'm energy efficient too!"

6. **Exit Through the Quantum Shop:** On their way out, guests pass through a gift shop selling quantum gadgets and souvenirs, like entangled pair socks (guaranteed to never lose their partner) and quantum encryption kits for kids. A puzzled customer asks, "Do you have anything in classical physics?" to which the cashier, wearing a t-shirt that says "Quantum or Bust!", replies, "You're in the wrong century for that!"

The infographic closes with Professor Quantum waving goodbye, standing next to a sign that reads, "Quantum Land: Where today's science fiction is tomorrow's science fact!" A small footnote adds, "Don't forget to check your quantum state before leaving. We're not responsible for any superpositions!"

Chapter 17

5G and beyond

Securing the future of connectivity

INTRODUCTION

- **A New Era of Connectivity:** How 5G and 6G technologies are revolutionizing communication.
- **Opportunities and Risks:** The benefits and potential threats associated with these advanced networks.
- **The Urgency of Security:** Why securing 5G and 6G is critical for the future of digital infrastructure.

Imagine a world where devices are seamlessly connected, data flows at unprecedented speeds, and smart cities operate with unparalleled efficiency. This is the promise of 5G and the forthcoming 6G technologies. However, as with any technological leap, these advancements bring new security challenges that must be addressed to safeguard the digital landscape.

THE THREAT LANDSCAPE OF 5G AND 6G

Increased Attack Surface: With the proliferation of connected devices, the attack surface expands significantly, providing more entry points for cyber threats.

Complex Network Architectures: The decentralized and software-defined nature of 5G and 6G networks introduces complexities that can be exploited by attackers.

Critical Infrastructure at Risk: The integration of 5G and 6G into critical infrastructure (e.g., healthcare, transportation, energy) amplifies the potential impact of cyber-attacks.

Real-World Examples

- **Example 1:** In 2020, the UK's National Cyber Security Centre identified vulnerabilities in Huawei's 5G equipment that could potentially be exploited by attackers.
- **Example 2:** The Mirai botnet attack on Internet of Things (IoT) devices demonstrated how easily connected devices could be co-opted into large-scale Distributed Denial of Service (DDoS) attacks, a risk that will only increase with 5G's proliferation.

KEY SECURITY CONCERNS

Data Privacy: The massive amounts of data transmitted over 5G and 6G networks raise significant privacy concerns. Ensuring data is encrypted and securely transmitted is paramount.

DOI: 10.1201/9781003562184-22

Network Slicing Vulnerabilities: 5G enables network slicing, which creates virtual networks tailored to specific applications. However, if not properly secured, these slices can become targets for attackers.

Supply Chain Security: The hardware and software components that make up 5G and 6G networks often come from various suppliers, increasing the risk of supply chain attacks.

IoT Security: The explosion of IoT devices connected to 5G networks introduces numerous vulnerabilities due to weak security measures in many IoT products.

MITIGATION STRATEGIES

Robust Encryption: Implement end-to-end encryption for data transmitted over 5G and 6G networks to protect against interception and eavesdropping.

Network Slicing Security (Protecting the Slices): Ensure that network slices are isolated from each other and implement strong access controls to prevent unauthorized access.

Network slicing in 5G allows for the creation of multiple virtual networks on a single physical infrastructure. Here's how to secure these slices.

Technical Implementation: How Network Slicing Works Software-defined networking (SDN) dynamically configures network slices, while network function virtualization (NFV) allows for flexible deployment of virtualized network functions.

Security Protocols: Safeguarding the Slices Isolation mechanisms like VLANs and firewalls separate slices. Strong access controls and encryption are essential for each slice, and intrusion detection systems (IDS) monitor for suspicious activity within slices.

Potential Vulnerabilities: The Cracks in the Armor Misconfigurations in SDN controllers or vulnerabilities in NFV infrastructure can compromise slice security. Resource exhaustion attacks pose risks that could impact multiple slices.

Supply Chain Risk Management: Conduct thorough vetting of suppliers and continuously monitor for vulnerabilities in both hardware and software components.

IoT Device Security: Enforce strict security standards for IoT devices, including regular firmware updates, strong authentication mechanisms, and network segmentation.

BEYOND MITIGATION: EMERGING SECURITY SOLUTIONS

Blockchain Technology: Explore how blockchain can be used to enhance security in 5G and 6G networks by providing a decentralized and tamper-proof method of data integrity and transaction verification.

Quantum-Resistant Cryptography: As quantum computing advances, traditional encryption methods may become obsolete. Implementing quantum-resistant cryptographic algorithms now can future-proof the security of 5G and 6G networks.

THE ROLE OF ARTIFICIAL INTELLIGENCE

AI-Driven Threat Detection: Use machine learning algorithms to detect anomalies and potential threats in real time, enhancing the ability to respond to attacks swiftly.

Automated Incident Response: Integrate artificial intelligence (AI) to automate responses to detected intrusions, dynamically altering the deception environment to keep attackers engaged.

Predictive Analytics: Leverage AI to predict potential attack vectors based on historical data and current threat landscapes, allowing for proactive defense measures.

Balancing technical and non-technical aspects

Practical Takeaways for Non-Technical Readers

- **Clear Explanations:** Provide straightforward explanations of technical concepts like network slicing and encryption.
- **Actionable Advice:** Offer practical steps that organizations can take to improve their 5G and 6G security posture without needing deep technical expertise.
- **Visual Aids:** Use diagrams and flowcharts to illustrate complex ideas, making them more accessible to a broader audience.

HUMAN ELEMENT IN 5G AND 6G SECURITY

User Training and Awareness: Develop comprehensive training programs to educate users on the security risks associated with 5G and 6G. Topics should include recognizing phishing attempts, the importance of strong passwords, and safe usage of IoT devices.

Cultural Shift: Promote a security-first culture within organizations where every employee understands their role in maintaining network security.

Geopolitical considerations

Cyberwarfare and Espionage: Discuss the potential for cyberwarfare and espionage targeting 5G and 6G infrastructure, particularly given the geopolitical tensions surrounding the development and deployment of these technologies.

National Security: Explore how different countries are approaching the security of their 5G and 6G infrastructure and the implications for global cybersecurity.

ETHICAL CONSIDERATIONS

Surveillance Concerns: As 5G and 6G networks become more integrated into critical infrastructure, there are significant concerns about the potential for increased surveillance. Discuss the ethical implications of data collection and how to balance security with privacy rights.

Transparency and Accountability: Emphasize the importance of transparency in data collection practices and holding entities accountable for the ethical use of data.

LOOKING AHEAD: THE FUTURE OF 6G

What Is 6G?: An overview of the upcoming 6G technology, its potential capabilities, and the anticipated timeline for its deployment.

Security Challenges Unique to 6G: Discuss the specific security challenges expected with 6G, such as higher data rates, increased device density, and new use cases like holographic communication.

Preparing for 6G: Steps organizations can take now to prepare for the security implications of 6G, including investing in research and development and enhancing current security infrastructures.

FOOD FOR THOUGHT

Balancing Security and Privacy: How can we strike a balance between national security and individual privacy in the age of 5G and 6G?

International Collaboration: What is the role of international collaboration in developing and enforcing security standards for next-generation networks?

Human Element: As technology continues to evolve, how can we ensure that the human element of cybersecurity remains a top priority?

THE DEMOCRATIZATION OF CYBERATTACKS

Attack-as-a-Service Platforms: Discuss how the rise of "attack-as-a-service" platforms is lowering the barrier to entry for launching sophisticated cyberattacks, even for individuals with limited technical expertise.

Raising Awareness: Explore the importance of raising awareness about these trends and encouraging the responsible development and use of security tools.

CONVERGENCE OF PHYSICAL AND DIGITAL SECURITY

Interconnected Security: With 5G and 6G enabling the IoT to flourish, the lines between physical and digital security are blurring.

Critical Infrastructure Risks: Discuss the implications of this convergence, such as the potential for cyberattacks to disrupt critical infrastructure controlled by IoT devices (e.g., power grids, transportation systems).

Evolving Security Strategies: Explore how security strategies need to evolve to encompass both physical and digital security measures.

THE ROLE OF REGULATION

Government Regulation: As 5G and 6G technologies develop, there will be ongoing debates about the role of regulation in ensuring security.

Benefits and Drawbacks: Discuss the potential benefits and drawbacks of government regulations on areas like equipment manufacturing, network infrastructure, and data privacy.

International Collaboration: Explore the possibility of international collaboration on developing and enforcing security standards for next-generation networks.

THE SECURITY IMPLICATIONS OF AI

Manipulating AI Systems: While AI plays a crucial role in 5G and 6G security through threat detection and incident response, there are also potential security risks associated with AI itself.

Sophisticated Attacks: Discuss the possibility of attackers manipulating AI systems or using AI to launch more sophisticated attacks.

Ensuring Trustworthiness: Explore the need for robust security measures to ensure the trustworthiness and reliability of AI used in 5G and 6G networks.

THE HUMAN ELEMENT AND THE FUTURE OF WORK

Distributed Workforce: As 5G and 6G empower automation and remote work models, explore the potential security concerns associated with a more distributed workforce.

Security Awareness Training: Discuss the need for security awareness training to encompass new work styles and the importance of securing home networks and devices used for remote work.

AUTONOMOUS SECURITY SYSTEMS

Self-Healing Networks: As 5G and 6G networks become more complex, autonomous security systems that can identify and mitigate threats without human intervention will be crucial. Discuss how self-healing networks, powered by AI, can detect anomalies, isolate affected nodes, and reconfigure themselves to maintain operational integrity.

Adaptive Security Models: Explore the potential for security systems that adapt in real time based on the evolving threat landscape. These systems could utilize machine learning to continuously learn from new threats and adjust their defense mechanisms accordingly.

PRIVACY-ENHANCING TECHNOLOGIES (PETS)

Differential Privacy: Introduce the concept of differential privacy, which adds statistical noise to data sets, allowing organizations to analyze data without compromising individual privacy. Discuss how this can be applied to data transmitted over 5G and 6G networks to protect user privacy while still enabling valuable insights.

Federated Learning: Explore the use of federated learning, where machine learning models are trained across multiple decentralized devices without exchanging data. This technique can enhance privacy and security by keeping sensitive data localized while still benefiting from collective learning.

GREEN CYBERSECURITY

Sustainable Security Practices: As environmental concerns grow, the energy consumption of 5G and 6G networks will come under scrutiny. Discuss the importance of developing security solutions that are not only effective but also energy efficient. Highlight the role of green cybersecurity practices in reducing the carbon footprint of these networks.

Energy Harvesting for IoT Security: Explore innovative methods for powering IoT devices using energy harvesting techniques (e.g., solar, kinetic). Discuss how these methods can ensure continuous operation of security measures in remote or hard-to-reach locations.

QUANTUM COMMUNICATION NETWORKS

Quantum Key Distribution (QKD): Delve into the potential of quantum communication networks for 5G and 6G, particularly focusing on QKD. QKD can provide theoretically unbreakable encryption, ensuring secure communication channels.

Post-Quantum Cryptography: With quantum computing on the horizon, discuss the development of cryptographic algorithms that can withstand quantum attacks. Explore how post-quantum cryptography will be essential for securing next-generation networks.

SOCIO-TECHNICAL SECURITY

Behavioral Biometrics: Introduce behavioral biometrics as a means to enhance security. This involves analyzing patterns in user behavior (e.g., typing speed, mouse movements) to create unique behavioral profiles that can help in authenticating users and detecting anomalies.

Human-Centric AI: Explore the integration of human-centric AI that focuses on understanding and predicting human behavior to improve security measures. Discuss how this approach can enhance the detection of social engineering attacks and improve user training programs.

THE EDGE COMPUTING PARADIGM

Decentralized Security: As edge computing becomes more prevalent with 5G and 6G, discuss the shift toward decentralized security models. Edge computing moves data processing closer to the source, reducing latency but also changing the security landscape. Explore how security can be maintained in a decentralized environment.

Zero Trust Architecture at the Edge: Implementing zero trust principles in edge computing environments, where every device, user, and network flow is authenticated and continuously validated, can significantly enhance security.

ETHICAL AI IN CYBERSECURITY

Explainable AI: Emphasize the importance of explainable AI in cybersecurity. As AI systems become more integrated into security operations, ensuring that their decision-making processes are transparent and understandable to humans is crucial for maintaining trust and accountability.

Bias Mitigation: Discuss strategies for mitigating biases in AI algorithms used for security purposes. Biased algorithms can lead to unfair treatment and missed threats. Highlight the importance of diverse data sets and continuous monitoring to reduce bias.

CONCLUSION

The advent of 5G and the impending arrival of 6G technologies present both unprecedented opportunities and significant security challenges. By understanding these challenges and implementing robust mitigation strategies, we can secure the future of connectivity and ensure that the benefits of these advanced networks are fully realized without compromising security.

CYBER CHUCKLES

- **Why did the hacker get a job in telecom?** They heard 5G has a lot of "bandwidth" for new opportunities!
- **Why don't cybersecurity experts like parties on 5G networks?** Too many guests "phishing" for connections!
- **What's a hacker's favorite type of wave?** A "data wave" – but they're always surfing illegally!
- **How did the 6G device feel at the security conference?** Out of its bandwidth!
- **Why did the IoT device start a music band?** It wanted to improve its "signal!"
- **What did the 5G tower say to the 6G tower?** "I'm all about the present, but you're the future wave!"

Part V

Specialized areas and practical implementations

Chapter 18

Operational technology (OT) security

Spoiler Alert!

- **Defense-in-Depth:** Learn how to implement multiple layers of security to protect critical OT systems.
- **Real-World Incidents:** Understand the impact of major OT security breaches like Stuxnet and Triton.
- **Proactive Measures:** Discover strategies for continuous threat intelligence and predictive security.
- **Unexpected Angles:** Explore innovative approaches like digital twins, gamification, and geopolitical considerations.

INTRODUCTION

Operational technology (OT) systems are integral to managing and controlling physical processes in various critical sectors, including manufacturing, energy, and transportation. As these systems increasingly converge with IT networks, they become more vulnerable to cyber threats. Effective OT security is essential not only to protect the infrastructure but also to ensure the continuity and safety of operations. This chapter explores the key strategies and best practices for securing OT environments, emphasizing the importance of a comprehensive approach that includes both strategic and operational measures.

HIGH-LEVEL OVERVIEW OF OT SECURITY

Strategic measures

1. **Visibility across OT and IT Environments**
   - Achieving visibility across both OT and IT networks is crucial for making informed security decisions. Comprehensive visibility allows for the identification of potential threats and vulnerabilities in real time, enabling proactive responses.
2. **Alignment between IT and OT Stakeholders**
   - Ensuring that IT and OT stakeholders are aligned on security posture and approach is vital. This alignment facilitates seamless communication and cooperation, leading to a more cohesive security strategy that addresses the unique needs of both environments.
3. **Justifying ROI for Security Investments**
   - Calculating and justifying the return on investment (ROI) for security measures is essential for gaining support from executive leadership. Demonstrating the financial and operational benefits of security investments helps ensure continued funding and prioritization of security initiatives.

DOI: 10.1201/9781003562184-24

4. **Intelligence-Driven Security Decisions**
   - Gathering the necessary intelligence to make security decisions that align with business objectives is key. This involves continuous threat intelligence and detection of anomalies, vulnerabilities, and zero-day threats across the IT and OT environments.

Operational measures

1. **Clear Inventory of Assets**
   - Maintaining a clear inventory of all assets and supporting infrastructure within the scope of operations is foundational for effective security management. This inventory should be regularly updated to reflect any changes in the environment.
2. **Properly Segmented Security Architecture**
   - Implementing a segmented security architecture that provides higher operational uptime and reduces the spread of threats is critical. Segmentation helps isolate potential breaches and limits their impact on the broader network.
3. **Continuous Threat Intelligence and Detection**
   - Employing continuous threat intelligence and detection mechanisms is essential for identifying and mitigating anomalies, vulnerabilities, and zero-day threats. This proactive approach helps in early detection and rapid response to potential security incidents.
4. **Response and Recovery Capabilities**
   - Developing the ability to respond to and recover from threats and attacks quickly and effectively is crucial for minimizing downtime and ensuring operational continuity. This includes having a robust incident response plan and regularly testing recovery procedures.

Immediate actions for enhanced OT security

1. **Whitelisting of Applications**
   - Implement application whitelisting to control which applications can run on OT systems, reducing the risk of malware and unauthorized software.
2. **Secure Remote Access Solutions**
   - Establish secure remote access processes and solutions, such as rendezvous servers, to protect remote connections to OT systems.
3. **Prohibition of Office Applications in Production**
   - Enforce policies that prohibit the use of office applications on production line systems to minimize the risk of malware infections and operational disruptions.
4. **Firewalling and Network Segmentation**
   - Strengthen firewall rules, especially egress filtering, and implement network segmentation to isolate critical OT systems from less secure parts of the network.
5. **Business Continuity Management (BCM) and Contingency Plans**
   - Develop and regularly update BCM and contingency plans to ensure that the organization is prepared to handle and recover from security incidents effectively.

DETAILED IMPLEMENTATION OF OT SECURITY

Integrating the four levels of OT security provides a comprehensive and structured approach to enhance the security posture of OT environments. These levels offer a step-by-step framework to address security challenges at varying degrees of complexity and integration.

Level 1: Foundational security

Asset Management

- **Inventory Control:** Develop a detailed inventory of all OT assets, including hardware, software, and firmware versions.
- **Regular Updates:** Maintain and update the inventory regularly to reflect changes and new deployments.

Basic Network Segmentation

- **VLAN Implementation:** Use Virtual Local Area Network (VLANs) to segment the network logically, separating critical OT systems from other network components.
- **Physical Separation:** Where possible, implement physical separation of critical assets to enhance security.

Access Control

- **User Authentication:** Implement strong authentication mechanisms for all users accessing OT systems.
- **Least Privilege:** Apply the principle of least privilege to minimize unnecessary access to critical systems.

Security Awareness Training

- **Regular Training:** Conduct regular security awareness training sessions for all employees, emphasizing the importance of OT security.
- **Phishing Simulations:** Perform phishing simulations to educate employees on identifying and responding to social engineering attacks.

Level 2: Intermediate security

Advanced Network Segmentation

- **DMZ Implementation:** Establish demilitarized zones (DMZs) to further isolate OT systems from IT networks.
- **Firewall Rules:** Implement advanced firewall rules to control and monitor traffic between segmented networks.

Intrusion Detection and Prevention

- **IDS/IPS Deployment:** Deploy Intrusion Detection Systems (IDS) and Intrusion Prevention Systems (IPS) tailored to OT environments.
- **Anomaly Detection:** Use anomaly detection systems to identify deviations from normal operational behavior.
- **IDPS in OT Environments:** Intrusion Detection and Prevention Systems (IDPS) are crucial for OT security. Let's explore their unique considerations in OT environments:
    - **OT-Specific Challenges: Unique Considerations** Deep packet inspection is needed for industrial protocols like Modbus and OPC UA. Identifying anomalies is challenging due to the variability of normal operations.
    - **Signature-Based vs. Anomaly-Based Detection: Choosing the Right Tool** Each approach has its strengths and weaknesses. Signature-based detection is effective for known threats, while anomaly-based detection identifies new, unknown threats. Examples of specific IDPS products suited for OT environments can provide more context.

- **Integration with SIEM: A Holistic View of Security** Integrating IDPS with Security Information and Event Management (SIEM) systems provides a centralized view of security events across IT and OT networks. Correlating events from different sources helps identify complex attacks.

Secure Remote Access

- **Multi-Factor Authentication (MFA):** Implement MFA for all remote access to OT systems.
- **Secure VPNs:** Use secure Virtual Private Networks (VPNs) to protect remote connections.

Patch Management

- **Regular Patching:** Develop a patch management strategy that prioritizes critical updates while minimizing operational disruptions.
- **Vendor Collaboration:** Work closely with OT vendors to ensure timely updates and security patches.

Level 3: Advanced security

Continuous Monitoring

- **SIEM Integration:** Integrate SIEM systems to centralize monitoring and analysis.
- **Real-Time Alerts:** Implement real-time alerting mechanisms to quickly identify and respond to potential threats.

Incident Response Planning

- **Tailored Playbooks:** Develop incident response playbooks tailored to specific OT scenarios.
- **Regular Drills:** Conduct regular incident response drills to ensure preparedness and identify areas for improvement.

Vulnerability Management

- **Regular Scanning:** Conduct regular vulnerability scans specifically designed for OT environments.
- **Risk Assessment:** Perform risk assessments to prioritize remediation efforts based on the criticality of vulnerabilities.

Regulatory Compliance

- **Industry Standards:** Ensure compliance with industry standards such as IEC 62443 and ISO 27001.
- **Regular Audits:** Perform regular audits of OT security practices to maintain compliance and identify areas for improvement.

Level 4: Optimized security

AI and Machine Learning

- **Behavioral Analysis:** Use AI and machine learning to analyze behavior patterns and detect anomalies in real time.

- **Predictive Security:** Implement predictive security measures that anticipate potential threats based on historical data and trends.

Threat Intelligence Sharing

- **Collaborative Networks:** Participate in threat intelligence sharing networks to stay informed about the latest threats and vulnerabilities.
- **Proactive Defense:** Use threat intelligence to inform proactive defense strategies and mitigate risks before they materialize.

Advanced Physical Security

- **Surveillance Systems:** Enhance physical security with advanced surveillance systems and access controls.
- **Environmental Controls:** Implement environmental controls to protect critical infrastructure from physical threats.

Business Continuity and Resilience

- **BCM Integration:** Integrate business continuity management with OT security to ensure operational resilience.
- **Recovery Planning:** Develop and test recovery plans to ensure quick restoration of operations following a security incident.

REAL-WORLD EXAMPLES OF OT SECURITY INCIDENTS

1. Stuxnet Worm
    - **Overview:** A sophisticated cyberattack targeting Iran's nuclear facilities, causing physical damage to centrifuges by exploiting vulnerabilities in Siemens Programmable Logic Controller (PLCs).
    - **Impact:** Highlighted the potential for cyberattacks to cause physical damage and disrupt critical infrastructure.
2. **Triton Malware**
    - **Overview:** A malware attack on a petrochemical plant's safety systems, designed to disable safety controls and cause catastrophic failures.
    - **Impact:** Emphasized the importance of securing safety systems and the potential consequences of compromised OT environments.
3. **Ukraine Power Grid Attack**
    - **Overview:** A cyberattack on Ukraine's power grid that resulted in widespread power outages, demonstrating the vulnerability of critical infrastructure to cyber threats.
    - **Impact:** Showcased the potential for cyberattacks to disrupt essential services and impact national security.

STANDARDS AND FRAMEWORKS FOR OT SECURITY

1. **ISA99/IEC 62443**
    - **Purpose:** Provides a comprehensive set of guidelines and best practices for securing industrial control systems.
    - **Overview:** Covers various aspects of OT security, including risk assessment, access control, and incident response.

2. NIST SP 800-82
   - **Purpose:** Offers guidelines for securing industrial control systems and integrating them with IT networks.
   - **Overview:** Provides a risk-based approach to OT security, emphasizing the importance of continuous monitoring and incident response.

THE EVOLVING THREAT LANDSCAPE FOR OT SECURITY

1. Targeted Attacks
   - **Trend:** Increasingly sophisticated cyberattacks specifically targeting OT environments.
   - **Example:** Advanced persistent threats (APTs) targeting critical infrastructure for espionage or sabotage.
2. Ransomware
   - **Trend:** Ransomware attacks extending to OT environments, disrupting operations and demanding ransom for system restoration.
   - **Example:** Ransomware attacks on manufacturing plants, halting production and causing significant financial losses.
3. Industrial Internet of Things IoT (IIoT) Security
   - **Trend:** The growing adoption of IIoT devices introduces new vulnerabilities and attack vectors.
   - **Example:** Compromised IIoT sensors leading to incorrect data inputs and potential operational failures.

SECURING OT SYSTEMS: BEST PRACTICES

1. Defense-in-Depth
   - **Approach:** Implement multiple layers of security to protect OT systems from various threats.
   - **Elements:** Include physical security, network segmentation, monitoring, access controls, and incident response.
2. Network Segmentation
   - **Strategy:** Divide the network into segments to contain potential breaches and limit lateral movement.
   - **Implementation:** Use VLANs, firewalls, and DMZs to isolate critical OT systems.
3. Access Controls
   - **Principle:** Apply the principle of least privilege to minimize unnecessary access to sensitive systems.
   - **Tools:** Use strong authentication, role-based access control (RBAC), and multifactor authentication (MFA).
4. Patch Management
   - **Practice:** Regularly update and patch OT systems to protect against known vulnerabilities.
   - **Challenges:** Balance the need for security updates with the requirement for operational uptime.
5. Vulnerability Assessments
   - **Process:** Conduct regular assessments to identify and remediate vulnerabilities in OT systems.
   - **Tools:** Use specialized tools and techniques tailored to OT environments.

THE HUMAN ELEMENT IN OT SECURITY

1. User Awareness Training
   - **Importance:** Educate OT personnel on security best practices and the latest threats.
   - **Methods:** Conduct regular training sessions, simulations, and phishing tests.
2. Insider Threats
   - **Risk:** Recognize that insider threats can pose significant risks to OT security.
   - **Mitigation:** Implement monitoring, access controls, and a culture of security awareness to mitigate insider threats.

THE FUTURE OF OT SECURITY

1. **Industrial IoT (IIoT) Security**
   - **Trend:** The integration of IIoT devices will require enhanced security measures to protect against new vulnerabilities.
   - **Approach:** Develop and implement security standards specifically for IIoT environments.
2. **Zero-Trust Architectures**
   - **Trend:** Adopting zero-trust principles to ensure that every access request is verified, regardless of its origin.
   - **Implementation:** Use continuous monitoring, identity verification, and strict access controls.
3. **Continuous Improvement**
   - **Principle:** OT security practices must evolve to keep pace with emerging threats and technological advancements.
   - **Strategy:** Regularly review and update security policies, conduct audits, and invest in ongoing training and education.

NEWER ANGLES ON OT SECURITY

1. **OT Security through the Lens of Supply Chain Risk**
   - **Concept:** OT security should be viewed as a vulnerability across the entire supply chain, not just within a single facility. A compromised supplier's system can create a domino effect, impacting multiple organizations down the line.
   - **Impact:** Highlighting the interconnected nature of supply chains underscores the potential for widespread disruptions if one link is compromised.
   - **Solutions:** Advocate for collaborative security measures throughout the supply chain, including shared threat intelligence and regular vulnerability assessments. This cooperative approach helps ensure that all parties are aligned and prepared to tackle potential threats.
2. **OT Security and the Rise of Digital Twins**
   - **Intrigue:** Digital twins are virtual replicas of physical systems that can be used for various purposes, including security.
   - **Explanation:** By using digital twins to simulate cyberattacks, organizations can identify potential vulnerabilities in OT environments before they are exploited. These simulations allow for proactive adjustments and improvements to security protocols.
   - **Benefits:** The proactive and preventative nature of using digital twins enhances overall security posture, reducing the likelihood of successful attacks and minimizing potential damage.

3. **Gamifying OT Security Awareness**
   - **Engagement:** Traditional security training can be dry and unengaging. Gamification introduces elements of fun and competition to the training process, making it more interactive and effective.
   - **Imagine:** Employees participate in simulations where they make security decisions in virtual scenarios. They learn from both their successes and failures, gaining practical insights into best practices.
   - **Benefits:** Gamification improves knowledge retention and encourages a more security-conscious workforce. It fosters a culture of continuous learning and vigilance, essential for maintaining robust OT security.
4. **OT Security and the Insurance Landscape**
   - **Shifting Landscape:** The rising costs of OT security incidents are impacting the cyber insurance industry, leading to changes in how policies are structured and priced.
   - **Incentives:** Insurance companies may start offering differential premiums or coverage limitations based on an organization's security posture. Those with stronger security measures might benefit from lower premiums.
   - **Collaboration:** Propose collaboration between OT security experts and insurance companies to develop more robust risk assessment and mitigation strategies. This partnership can lead to better-informed policies that incentivize good security practices.
5. **OT Security and Geopolitical Tensions**
   - **New Threat Actors:** Nation-state actors increasingly target OT infrastructure for espionage or sabotage, particularly in the context of geopolitical tensions.
   - **Mitigation Strategies:** Increased government regulations, international cooperation, and information sharing are essential to combat these sophisticated threats. Public-private partnerships can also play a crucial role in enhancing overall security.
   - **Ethical Considerations:** The ethical implications of offensive cyberwarfare must be considered. Responsible disclosure of vulnerabilities and adherence to international norms and laws are crucial to maintaining global cybersecurity stability.

CONCLUSION

As OT systems become more interconnected with IT networks, the need for robust OT security measures grows increasingly urgent. By understanding the unique challenges of OT security and implementing a comprehensive framework that includes the four levels of security, organizations can protect their critical infrastructure and ensure the safe and reliable operation of industrial processes. A holistic approach to OT security, encompassing risk assessment, network segmentation, monitoring, patch management, employee training, and incident response, is essential for mitigating cyber threats and safeguarding the future of industrial operations.

Integrating unexpected angles, such as viewing OT security through the lens of supply chain risk, leveraging digital twins for proactive security, gamifying security awareness, collaborating with the insurance industry, and considering geopolitical tensions, adds depth and breadth to the security strategy. These innovative approaches highlight the importance of staying ahead of emerging threats and continuously improving security measures.

By embracing these strategies, organizations can create a resilient OT security framework that addresses the complexities of modern industrial environments. The final chapter emphasizes the importance of a proactive and comprehensive approach to OT security, ensuring the protection of vital systems and processes in an increasingly interconnected world.

CYBER CHUCKLES: OPERATIONAL TECHNOLOGY EDITION

Why did the OT engineer bring a ladder to work?

Because he wanted to reach the top level of security!

Knock, knock.

Who's there?

Anomaly detection.

Anomaly detection who?

Just checking if everything's normal around here.

What do you call a sophisticated cyberattack on a power grid?

A shocker!

How many OT security professionals does it take to change a light bulb?

None. They just need to ensure the system is resilient enough to handle it automatically.

Why don't OT systems trust each other?

Because they know better than to let their guard down in a high-risk environment.

What do you get when you cross a ransomware attack with a factory?

A production of "The Great Lockdown!"

Why did the cybersecurity expert go broke?

Because he couldn't find any vulnerabilities to exploit for extra cash.

Why are OT networks like onions?

Because they have layers, and sometimes they make you cry when things go wrong.

Did you hear about the hacker who tried to break into a water treatment plant?

He got flushed out pretty quickly!

Why did the factory robot apply for a cybersecurity job?

Because it wanted to keep its circuits in check!

How do OT systems stay in shape?

They do regular drills and exercises to stay strong!

Why was the control system always calm during an attack?

Because it had plenty of "resilience" training.

Chapter 19

Weaving the security fabric

Integrating remote work, emerging technologies, and edge computing

Spoiler Alert!

- **Zero-Trust Everywhere:** Discover how a zero-trust architecture with micro-segmentation and context-aware authentication forms the bedrock of secure remote work environments.
- **Tech and Beyond:** Explore how emerging technologies like blockchain and Augmented Reality (AR)/Virtual Reality (VR) reinforce security, while edge computing enhances real-time threat detection.
- **Futureproofing:** Dive into advanced concepts like quantum-resistant cryptography, federated learning, and neurosecurity to stay ahead of evolving cyber threats.

INTRODUCTION

Securing remote work environments, emerging technologies, and edge computing has never been more critical. With the rapid evolution of cyber threats, it's essential to stay ahead of the curve and protect our digital assets effectively.

Opening Hook: A Case in Point: Imagine Sarah, a remote worker for a major corporation. One day, she receives an email that looks like it's from her IT department, asking her to verify her credentials. Trusting the source, she clicks the link and enters her details. Within minutes, cybercriminals have accessed the company's network, leading to a $4 million loss. This incident underscores the urgent need for robust security measures in our increasingly digital world.

> Cyberattacks targeting remote workers have increased by 600% since 2019, illustrating the heightened vulnerability in distributed work environments.

Remote work as the foundation

The foundation of our security fabric is built on securing remote work environments, which have become the norm. This involves adopting a zero-trust architecture to ensure continuous verification of all users and devices.

ZERO-TRUST WITH A TWIST: MICRO-SEGMENTATION AND CONTEXT-AWARE AUTHENTICATION

- **Zero-Trust Architecture:** Zero-trust architecture forms the foundation of the security fabric. It treats no device or user as inherently trusted, requiring continuous verification in remote work environments. This approach ensures that every access request is authenticated and authorized, minimizing potential entry points for attackers.

DOI: 10.1201/9781003562184-25

- **Micro-Segmentation:** Go beyond basic multifactor authentication (MFA). Explore how micro-segmentation within the zero-trust architecture can further restrict lateral movement within the network if a breach occurs. This involves creating smaller, isolated network segments that limit the spread of an attack.
- **Context-Aware Authentication:** Introduce the concept of context-aware authentication that utilizes factors like user location, device type, and time of day to dynamically adjust authentication requirements. This adds an extra layer of security without unnecessary friction for legitimate access attempts.
- **Specific Frameworks:** Mention the "Principle of Least Privilege" to give readers a deeper understanding of zero-trust frameworks. This principle limits user access rights to the bare minimum required to perform their jobs, reducing the attack surface.

Best Practices for Zero-Trust Implementation: To solidify your understanding of zero-trust, consider these best practices:

- Implement MFA to add an extra layer of security.
- Use network segmentation to isolate critical systems and data.
- Continuously monitor user behavior and network activity for anomalies.

Transitioning to the role of emerging technologies, let's explore how these can reinforce our security fabric.

EMERGING TECHNOLOGIES AS REINFORCEMENTS

Emerging technologies like blockchain and AR/VR offer innovative ways to enhance security measures.

Blockchain: Highlight how blockchain technology can reinforce the security fabric by providing tamper-proof audit trails for remote access logs and data transactions. This enhances trust and accountability by ensuring that all actions are recorded in an immutable ledger.

- **Permissioned Blockchain Networks:** Explore permissioned blockchain networks as a potential solution for secure remote access logs. These networks offer more control and privacy compared to public blockchains, making them ideal for corporate environments.
- **Security Benefits**
  - Immutable ledgers that prevent tampering.
  - Decentralized nature reducing single points of failure.
- **Security Challenges**
  - **Initial Coin Offerings (ICOs):** Highlight the security issues associated with ICOs, such as fraudulent schemes and hacking incidents.
    - Red flags to watch out for: unrealistic guaranteed returns, anonymity of the development team.
  - **Rug Pulls in Decentralized Finance (DeFi):** Discuss the rise of "rug pulls" where developers abandon a project after collecting investor funds, leading to significant financial losses.
- **Best Practices**
  - Conduct thorough code audits for smart contracts.
  - Use established and secure blockchain platforms.
  - Implement multi-signature wallets to enhance security.

Transitioning from blockchain to AR/VR, these technologies also play a crucial role in our security fabric.

AR/VR: Discuss how AR/VR can be used to create immersive security training simulations for employees. These simulations can make employees more effective in identifying and mitigating cyber threats by providing realistic scenarios and hands-on practice.

- **Potential for Social Engineering Attacks:** Highlight how attackers can manipulate virtual environments to trick users into divulging sensitive information. For example, creating seemingly legitimate virtual stores to steal user payment information.
- **Data Breaches of User Profiles:** Highlight the risks of unauthorized access to user profiles containing personal and sensitive data.
- **Best Practices**
  - Implement end-to-end encryption for data transmitted within AR/VR applications.
  - Use biometric authentication to secure user access to AR/VR platforms.
  - Regularly update AR/VR software to patch vulnerabilities.
- **Quantum-Resistant Cryptography:** Explore the development of quantum-resistant cryptographic algorithms to protect data against potential future quantum computer attacks.
- **Homomorphic Encryption:** Discuss the potential for searchable encryption within the security fabric, allowing authorized users to search encrypted data without decrypting it.

Now, let's examine how edge computing adds distributed strength to our security fabric.

EDGE COMPUTING AS THE DISTRIBUTED STRENGTH

Edge computing enhances security by processing data closer to the source, reducing latency, and improving real-time threat detection.

Edge Devices as Sensors: Explain how edge devices in a remote work setup can act as distributed sensors within the security fabric. These devices collect real-time data on potential threats and suspicious activity, providing a continuous stream of information for security analysis.

- **Best Practices**
  - Implement strong authentication mechanisms for edge devices.
  - Use role-based access controls to limit device access to authorized users.
  - Encrypt data transmitted between edge devices and central servers.

AI-Powered Threat Analysis at the Edge: Explore the potential for artificial intelligence (AI) and machine learning to analyze data collected from edge devices in real time. This enables faster threat detection and response, allowing organizations to quickly mitigate potential security breaches.

- **Best Practices**
  - Deploy centralized monitoring tools to oversee edge device security.
  - Regularly update firmware and software on edge devices to address vulnerabilities.

Zero-Trust Networking at the Edge: Introduce the concept of zero-trust service meshes for secure communication between edge devices and cloud services.

Federated Learning for Threat Detection: Discuss the potential of federated learning for collaborative threat intelligence at the edge.

TECHNICAL DETAILS – A DEEP DIVE

For readers with a technical background, delve deeper into specific protocols and frameworks.

Protocols and Implementations: Discuss specific protocols like OAuth 2.0 for zero-trust environments, explaining their role in secure authentication and authorization. Delve into the inner workings of blockchain technology, including consensus mechanisms like Proof of Work (PoW) and Proof of Stake (PoS). Address the challenges and solutions for implementing AI-powered threat analysis at the edge, including data privacy concerns and computational limitations.

Zero-Trust Implementation Frameworks: Explore specific frameworks like the Zanzibar Model or the SPIRE Model for implementing zero-trust architecture.

Blockchain for Secure Boot: Delve into the concept of using blockchain technology to secure the boot process of edge devices.

Weaving it together – a scenario with a futuristic twist

To illustrate the integration of these technologies, let's look at a futuristic scenario.

AI-Powered Insider Threat Detection: Craft a scenario where an AI at the edge detects an anomaly in user behavior, potentially indicating an insider threat. The security fabric triggers MFA, micro-segmentation isolates the compromised device, and forensic analysis is initiated.

Potential Challenges and Solutions: Complexity of Managing Distributed Security Architecture: Discuss the challenges of integrating and managing various components of the security fabric across distributed environments.

- **Solution:** Advocate for centralized management platforms that provide visibility and control over the entire security infrastructure.

Need for Employee Training on New Technologies: Highlight the importance of training employees to effectively use and understand the new security technologies and practices.

- **Solution:** Implement continuous training programs and provide resources to keep employees updated on the latest security protocols and tools.

EMERGING TRENDS

Biometric Authentication at the Edge: Discuss the potential for on-device biometric authentication (fingerprint, facial recognition) for edge devices.

Neuromorphic Computing for Real-Time Threat Analysis: Explore the potential of neuromorphic computing chips for real-time threat analysis at the edge.

Neurosecurity

Emerging Field of Neurosecurity: Briefly touch on the emerging field of neurosecurity, which focuses on protecting the human brain from cyberattacks. While this may seem futuristic, exploring potential solutions like brain authentication can spark discussions about the future of human-computer interaction in the security landscape.

CONCLUSION – THE EVOLVING FABRIC

To wrap up, let's consider the future of our security fabric.

Continuous Adaptation: Emphasize the need for the security fabric to continuously adapt and evolve as new technologies emerge and cyber threats become more sophisticated.

Collaboration Is Key: Highlight the importance of collaboration between IT, security teams, and employees to maintain a strong and unified security fabric.

By weaving together the strengths of remote work security, emerging technologies, and edge computing, this chapter provides a comprehensive and cohesive narrative on creating a robust security ecosystem for the future of work. This approach ensures that all aspects of cybersecurity are integrated and working in harmony to protect the digital landscape.

CYBER CHUCKLES

1. **Remote Work Realities:** Why did the cybercriminal go after the remote worker? Because home is where the Wi-Fi is weakest!
2. **Blockchain Blunders:** Why don't blockchain developers tell secrets at parties? Because they can't find the right keys!
3. **AR/VR Adventures:** Why did the hacker enjoy VR meetings? Because it was the perfect place to phish in virtual waters!
4. **Edge Computing Edges:** Why did the edge device go to therapy? It had too many issues to process alone!
5. **Zero-Trust Zen:** Why did the zero-trust architect meditate? To keep his mind segmented and context-aware!
6. **Quantum Quandaries:** Why did the quantum computer fail its driving test? It couldn't stay in a single lane!

Chapter 20

The beautiful world of AI

Spoiler Alert!

- Imagine your toaster trying to get on your Wireless Fidelity (WIFI) password (because apparently, AI wants to browse cat videos too).
- This chapter explores the wild world of regulating AI before it takes over the world (or at least your Netflix queue).
- We learn how AI can be biased (think a loan app that discriminates against people who like to listen to polka music).
- The chapter warns us about AI black boxes: complex systems no one understands (because who wants a doctor who can't explain why they're prescribing robot crickets?).
- Get ready for tips on building a responsible AI like setting clear goals (like "help people, not steal their jobs") and choosing the right training data (because you don't want your AI to learn from conspiracy theories).
- The chapter wraps up with everyone from society to individuals taking responsibility for keeping AI in check (because with great power comes great meow-sponsibility).

INTRODUCTION

Today, artificial intelligence (AI) is pervasive, revolutionizing sectors from healthcare diagnostics to optimizing manufacturing operations, and reshaping financial crime detection. In our daily lives, AI intervenes in hiring, loan approvals, and even insurance industries. For instance, a machine learning system could analyze a job applicant's video interview, scrutinizing everything from micro-expressions to body temperature to aid in decision-making.

The AI market's projected growth, from US$119.78 billion in 2022 to an anticipated US$1597.1 billion by 2030, reflects its ubiquity. Thus, AI regulation is paramount for several reasons:

- **Impact on Society:** AI's reach extends to critical areas like healthcare and justice, necessitating responsible use to avert discrimination or privacy breaches.
- **Bias and Fairness:** AI could perpetuate biases if unchecked. Regulation ensures fairness and non-discrimination in AI systems.
- **Transparency:** The intricate nature of many AI systems demands clarity to bolster individual trust in the decision-making process.
- **Security:** With AI's potential for unintended outcomes, regulation safeguards critical domains, including aviation and military applications, ensuring safety and preventing harm.

DOI: 10.1201/9781003562184-26

THE MODEL

Creating an AI (governance) model (Figure 20.1) is akin to teaching a very literal genie how to grant wishes – you need to be incredibly specific about what you want, or you might end up with something wildly unexpected. At the heart of this process is a blend of legal, ethical, and logical considerations that ensure our digital genie doesn't run amok.

First, think about the application of the AI system as if deciding whether a teenager should have access to the family car. It's crucial to consider which applications should be given the keys to all the data in the AI system. Just like with the car, you wouldn't want it used for a purpose it wasn't intended for.

Then, comes the input stage, which is much like ensuring the teenager knows not just how to drive but also where they're going. The input to an AI system needs to be sanitized and aligned with the system's original goal, much like making sure the teenager knows the rules of the road and has a clear destination.

Defining a clear goal for the AI is like setting a curfew. It provides a framework within which the AI operates, ensuring that its actions are legal and logical. The data used is like the fuel for the car; it needs to be clean, legally obtained, and suitable for the journey ahead.

Choosing the right algorithm is like picking the right car for the trip. Just as you wouldn't use a two-seater sports car for a family holiday, you need to select the appropriate algorithm for the data and task at hand.

Training the AI system is akin to teaching the teenager to drive. It's time-consuming and requires patience. Monitoring the system's performance ensures it learns correctly and doesn't adopt any bad habits. It's also crucial to design this process carefully to avoid biases, much like ensuring the teenager doesn't learn to drive by watching racing movies.

Finally, when it comes to making decisions with the AI system, it's like deciding when the teenager can drive solo. The faster a decision is needed, the less human oversight there might be. However, ensuring that these decisions can be explained is like being able to trace back the teenager's driving route and choices, ensuring transparency and accountability.

In essence, guiding an AI model from conception to implementation is a delicate balance of clear goals, careful selection of data and algorithms, and constant oversight, all while maintaining transparency and accountability. Just like teaching a teenager to drive, it requires setting clear boundaries, providing the right tools and guidance, and trusting but verifying their actions.

When crafting a machine learning system, weaving through the maze of legal, ethical, and logical threads is paramount. Picture this: every AI model is essentially a miniaturized, digital mirror of our vast, unpredictable world, sculpted by training algorithms to dance to the rhythm of data (Figure 20.1).

Here's how to steer this complex ballet to a harmonious performance:

1. **Application Access**: First off, question if the application really needs the keys to the kingdom. Does it require full access to the AI system's data trove, or just a peek into the courtyard?
2. **Input Integrity**: Next, scrutinize the data morsels fed into your AI. Are they clean, relevant, and in line with the grand vision you've set for the system? Garbage in, garbage out, as they say.
3. **Defining Goals**: The compass for your journey – clearly define what you're aiming to achieve with your machine learning quest. This goal is your North Star, guiding the legal and logical pathways you'll navigate.

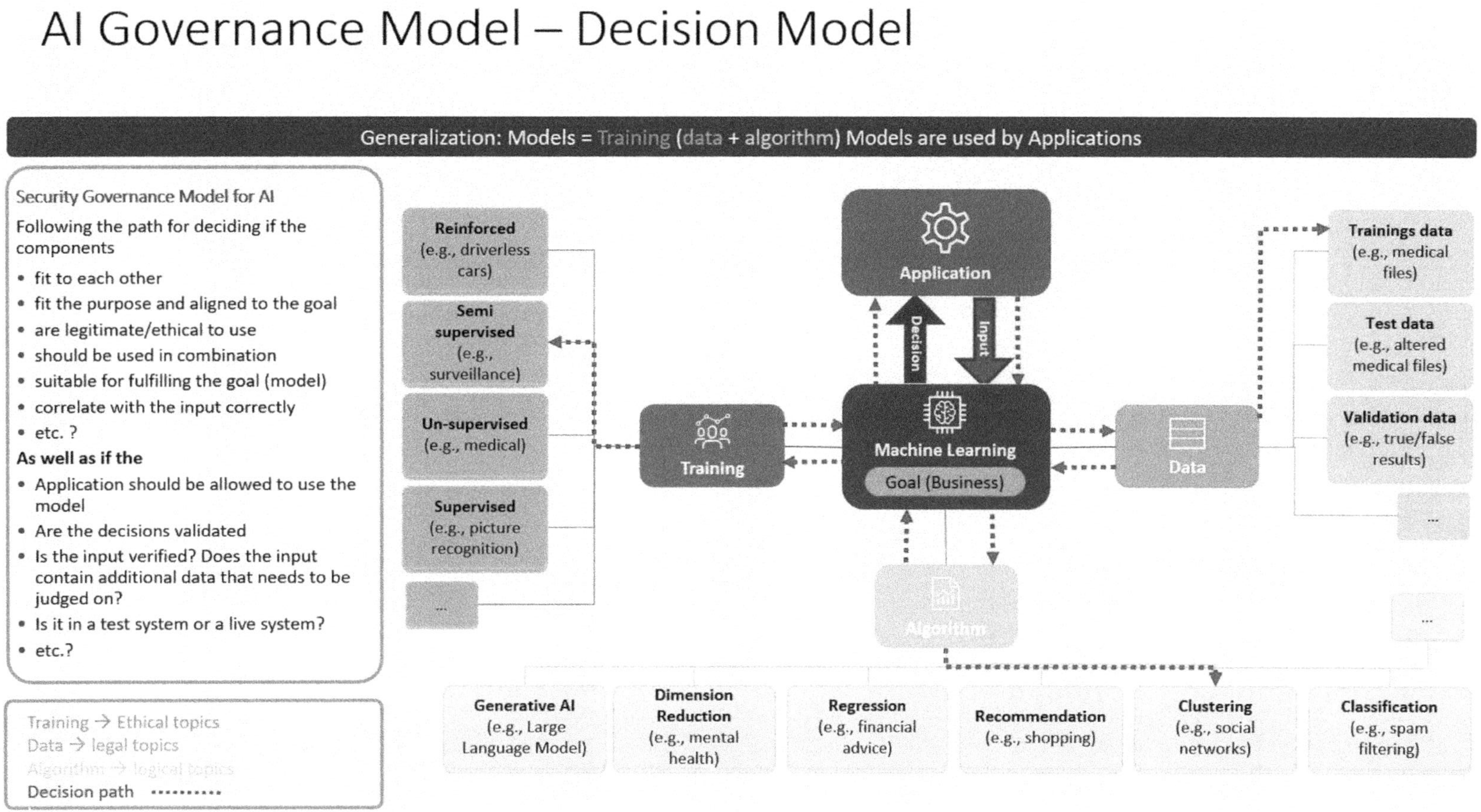

Figure 20.1 The AI Governance Decision Model in three major parts: data, algorithm, and training.

4. **Data Legality:** Dive into the data pool with caution. Every piece of information must be legally obtained and fit for the system's purpose. Think of it as curating ingredients for a Michelin-starred meal; quality and sourcing matter.
5. **Algorithm Selection:** Choosing the right algorithm is akin to selecting the perfect conductor for your orchestra. Each type of data and task requires a maestro to bring out the best performance, so pick wisely to hit the right notes.
6. **Mindful Training:** The training stage is a marathon, not a sprint. It's crucial to keep a vigilant eye on your AI's learning curve, ensuring it's absorbing the right lessons without falling into the traps of human biases or ethical slip-ups.
7. **Transparent Decisions:** Finally, when it's time for your AI to make decisions, blend human touch with machine efficiency. Speed may dictate the level of human oversight, but transparency is non-negotiable. Ensure that the curtain is pulled back on how decisions are made, illuminating the AI's thought process for all to see.

In essence, guiding an AI system from conception to decision-making is a delicate dance of blending technology with human insight. By following these steps, you create a pathway that ensures your AI not only performs its tasks with precision but does so in a manner that is ethical, legal, and, above all, transparent. This approach not only elevates the AI's performance but also builds trust in the technology, laying a solid foundation for its responsible use in society.

TRANSPARENCY: UNVEILING THE INNER WORKINGS OF AI

Establishing trust in the realm of AI isn't just about showcasing its capabilities; it's about opening the curtain to reveal the mechanisms behind the magic. Transparency is the cornerstone upon which trust is built, allowing stakeholders, from customers to employees, to understand, question, and ultimately have faith in the AI systems they interact with.

AI algorithms in action: Decision trees and neural networks

AI algorithms like decision trees and neural networks are powerful tools in cybersecurity. Here's how they work.

- **Decision Trees: Branching Out for Security** Decision trees are used for classification tasks in cybersecurity, such as identifying malicious emails or network traffic. Here's a simple example of a decision tree for classifying network traffic as benign or malicious.
- **Neural Networks: Learning the Patterns of Threats** Neural networks are used for anomaly detection and threat prediction in cybersecurity. Deep learning, a subset of neural networks, is applied to complex cybersecurity problems.
- **Bias Mitigation Techniques: Ensuring Fairness and Accuracy** Reducing bias in AI models involves techniques like data preprocessing, algorithmic adjustments, and fairness constraints. Biased AI models can lead to discriminatory outcomes in cybersecurity, so it's crucial to implement bias mitigation techniques.

Creating a window into AI's soul

To foster this transparency, it's imperative to illuminate each stage of the AI model's lifecycle. From the initial data input, through the intricacies of algorithm selection, to the nuances of the training process, every step should be as open as a book. Picture providing a behind-the-scenes tour of how the AI system works, what data nourishes its growth, and how it reaches its conclusions.

Demystifying the process

Imagine if every AI system came with its own "nutrition label," detailing its ingredients (data inputs), its dietary regimen (algorithm and training process), and its health check results (decision-making rationale). This approach demystifies the AI process, transforming it from a black box into a glass house where everything is visible and understandable.

The consequences of keeping secrets

Failing to maintain transparency can erode trust as quickly as sand through a sieve. Without clarity about how decisions are made or what data is used, stakeholders might question the AI system's integrity and reliability. It's akin to a magician who refuses to reveal the secret behind a trick, leaving the audience skeptical rather than amazed.

Questions are the key

Encouraging questions and providing comprehensive answers is the bedrock of establishing transparency. While the chapter provides a glimpse into potential inquiries, it's crucial to understand that these initial questions are just the tip of the iceberg. Stakeholders may have a myriad of concerns or curiosities, ranging from data privacy to ethical implications. Being prepared to address these questions not only enhances transparency but also solidifies trust in the system.

A transparent future

The path to responsible and ethical AI use is paved with transparency. By shedding light on the AI process, society, organizations, and individuals can foster an environment where AI systems are not only trusted but also held to high ethical and responsible standards. The collective responsibility to maintain this transparency ensures that AI contributes positively to society, adhering to our shared values and promoting the public good (Figure 20.2).

In the digital age where AI permeates every facet of decision-making, the cornerstone of fostering trust lies in the realm of transparency. Ensuring a clear view into the AI model's lifecycle – from data sourcing and algorithmic underpinnings to the final training stages – is not just beneficial, it's imperative. This open-book approach allows stakeholders, be they customers or team members, to grasp the inner workings and rationale behind AI-driven decisions, thereby fostering a foundation of trust.

Creating a culture of transparency involves demystifying each phase of the AI development process. It's about laying bare the journey from raw data to decision-making, ensuring that every step, from the selection of algorithms to the inputs that feed them, is clearly articulated and open to scrutiny. This transparency not only bridges the gap between AI systems and their users but also reinforces the integrity of the decision-making process itself.

The absence of transparency, on the other hand, risks undermining trust and damaging the very fabric of stakeholder relationships. It's crucial, therefore, for entities leveraging AI to adopt an ethos of openness, addressing any queries or concerns with candor. Imagine the potential questions as just the starting point for a deeper dialogue aimed at reinforcing transparency and cultivating trust.

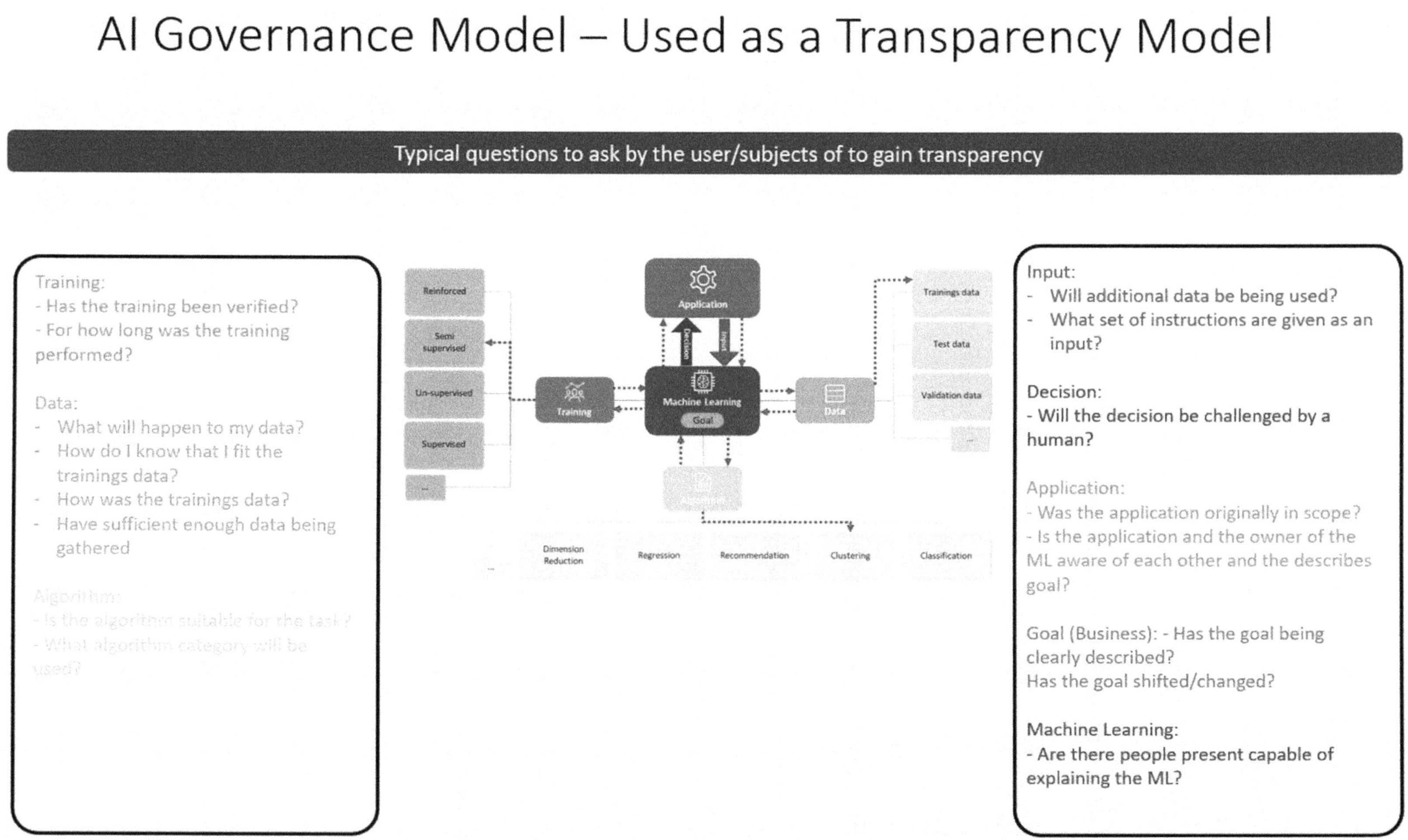

Figure 20.2 Utilizing the AI Governance Model for questioning purposes, for creating certainty and transparence.

CONCLUSION: A COLLECTIVE RESPONSIBILITY

The ethical deployment of AI transcends individual or organizational boundaries, placing the mantle of responsibility on society, organizations, and individuals alike.

- **Society's Role:** It falls upon the broader community to ensure that AI's evolution harmonizes with societal values and contributes positively to the common good. Through the establishment of regulatory frameworks and oversight mechanisms, society can guide the ethical development and application of AI technologies.
- **Organizational Accountability:** Entities at the helm of AI development and utilization bear a critical duty to ensure their creations are ethically grounded, transparent, and equitable. This includes a commitment to preventing the reinforcement of existing biases, ensuring AI decisions are understandable and justifiable, and safeguarding against any potential harm.
- **Individual Engagement:** On a personal level, there's an onus on everyone to become informed about the AI systems that increasingly influence our lives. Beyond mere awareness, there's a call to action for individuals to advocate for ethical AI practices and to voice concerns when AI's application strays from the path of responsibility and fairness.

In weaving transparency into the fabric of AI, we not only bolster trust but also pave the way for responsible and ethical AI utilization, ensuring technology serves as a force for good, aligned with the values and aspirations of all stakeholders involved.

CYBER CHUCKLES: NAVIGATING THE BRAVE NEW WORLD OF AI WITH A SMILE

Welcome to the future, where your toaster is sneakily trying to join your Wi-Fi network – not for malicious purposes, but to catch up on the latest cat videos. This is the world AI promises us: smart, interconnected, and maybe just a tad overzealous about pet content.

Regulate or be dominated

Before we find ourselves in heated debates with our refrigerators over electricity bills, there's a pressing need to put some reins on AI. Yes, it's about ensuring our Netflix queues remain untouched by machine learning algorithms with questionable taste in movies. But more importantly, it's about keeping the digital future bright and ethical, not just filled with unexpected movie recommendations.

Bias in the machine

AI, much like a misunderstood artist, can sometimes get it wrong. Picture a loan application system with a peculiar penchant for rejecting polka enthusiasts. Sounds absurd? Welcome to the world of AI bias. This chapter dives deep into the quirks of AI decision-making, reminding us that even machines can pick up some of our less admirable traits.

The black box dilemma

Ever tried to get a straight answer out of a mystic orb? That's what dealing with AI black boxes feels like. These complex systems, with their inscrutable decision-making processes, could one day recommend robot crickets for medical treatment without explaining why. It's time we demanded more transparency from our digital counterparts, lest we find ourselves at the mercy of an AI doctor with inscrutable prescriptions.

Crafting responsible AI

Building AI with a moral compass involves setting clear objectives like "assist humanity" rather than "plot world domination" and feeding it data that's as unbiased as possible. After all, we don't want AI to start believing the moon landing was a high-budget film shoot. This chapter equips you with the blueprint for nurturing AI systems that uplift rather than undercut, focusing on the benefits of selecting ethical training data and defining altruistic goals.

A united front against the AI apocalypse

It's on all of us, from society to individuals, to keep AI from straying into the realm of sci-fi horror. Whether it's demanding transparency from AI developers or ensuring AI-enhanced toasters don't start their own online shopping sprees, a collective effort is crucial. Remember, with great computational power comes great responsibility – let's make sure our AI future is filled with helpful assistants, not digital overlords with an affinity for cat videos.

In conclusion, as we stand on the cusp of a world brimming with AI potential, it's essential to steer this ship with a keen eye on ethics, transparency, and responsibility. By doing so, we ensure a future where AI enriches our lives, keeps our toasters in line, and maybe – just maybe – helps us understand why cats are so fascinated by boxes. Welcome to the beautiful, bewildering world of AI, where every day is an adventure in digital ethics.

Part VI

Holistic and forward-looking perspectives

Chapter 21

Rethinking access for AI

Spoiler Alert!

- Get ready for mind-bending concepts like the "DIKW pyramid" (data, information, knowledge, wisdom) and how AI needs to navigate its way through this pyramid to function effectively.
- The chapter wraps up with some comical advice on how to keep your AI overlords happy (and secure) in a world where the lines between machines and humans are blurring.

THE EVOLUTION OF ACCESS CONTROL IN THE AGE OF KNOWLEDGE

In the ever-evolving landscape of information security, the journey through the Data-Information-Knowledge-Wisdom (DIKW) pyramid reflects not just the progression of data processing but also the development of access control models over the years. This chapter delves into how traditional access control mechanisms like Mandatory Access Control (MAC) and Discretionary Access Control (DAC) laid the groundwork for securing data and information. Yet, as we ascend the pyramid toward knowledge, the complexities of managing access in a way that fosters insightful analysis and decision-making become apparent. It becomes crucial to adopt more dynamic and context-aware access control models such as Attribute-Based Access Control (ABAC) and context-aware access control to navigate the intricate landscape of knowledge generation effectively.

THE FOUNDATION OF ACCESS CONTROL: DATA AND INFORMATION LAYERS

At the base of the DIKW pyramid, the data layer represents the raw, structured input that organizations collect and store. Here, traditional access control models like MAC and DAC have historically provided the foundation for security. MAC, with its emphasis on classifications and central authority control, and DAC, which offers flexibility through owner discretion, have been instrumental in protecting this foundational layer from unauthorized access and breaches.

- **Mandatory Access Control** (MAC): This model, characterized by its stringent control and classification system, ensures that access to data is governed by a centralized policy, making it suitable for environments where confidentiality is paramount.
- **Discretionary Access Control** (DAC): DAC allows resource owners to manage access based on their discretion, offering a flexible approach that facilitates collaboration and data sharing, particularly valuable at the information layer where structured data begins to be analyzed and interpreted.

DOI: 10.1201/9781003562184-28

These models, while effective for managing access to data and structured information, begin to show limitations as organizations ascend the pyramid. The shift toward the information layer, where data is aggregated, analyzed, and turned into actionable insights, necessitates a more nuanced approach to access control. Role-Based Access Control (RBAC) emerges as a fitting model at this juncture, aligning access permissions with organizational roles and job functions, thereby streamlining the management of access rights in line with the evolving nature of the content.

TRANSITIONING TO KNOWLEDGE: THE NEED FOR DYNAMIC ACCESS CONTROL

As the focus shifts from structured information to the generation of knowledge, the limitations of traditional access control models become increasingly apparent. Knowledge generation, particularly in contexts like data lakes, requires a holistic view of data that is often incompatible with the compartmentalized access provided by MAC, DAC, and even RBAC. To generate accurate and comprehensive knowledge, there's a need for access control models that can accommodate the complex, situational nature of this endeavor.

- **Attribute-Based Access Control (ABAC):** ABAC's flexibility, which considers a wide array of user, resource, and environmental attributes, makes it exceptionally suited for the knowledge layer. This model enables nuanced access decisions that reflect the complexity of knowledge generation, allowing for more granular and context-sensitive control.
- **Context-Aware Access Control:** Building on the principles of ABAC, this model integrates real-time contextual information into access decisions. By considering factors such as user location, device security posture, and time of access, context-aware controls ensure that knowledge access is both secure and conducive to insight generation.

These dynamic and adaptive models represent a significant evolution in access control, tailored to the unique demands of the knowledge layer. They facilitate a level of access that is both broad to encourage comprehensive analysis and controlled to protect against unauthorized use or exposure of sensitive data.

THE IMPLICATIONS OF ADVANCED ACCESS CONTROL FOR CYBERSECURITY

As organizations climb the DIKW pyramid, reaching the knowledge layer necessitates a sophisticated approach to access control that not only secures sensitive data but also enables the fluid generation of insights. This balance is critical in maintaining robust cybersecurity while fostering an environment where knowledge can be derived from comprehensive data analysis. The introduction of ABAC and context-aware access control models marks a significant evolution in access management, tailored to meet these dual needs.

- **Ensuring Security in a Dynamic Data Environment:** ABAC offers a highly adaptable framework for access control, capable of accommodating the multifaceted nature of knowledge generation. By evaluating a broad spectrum of attributes – encompassing user characteristics, resource specifics, and environmental factors – ABAC ensures that access decisions are made with a nuanced understanding of each request's context. This granularity not only enhances security by limiting access to those with a legitimate need but also supports compliance by dynamically enforcing policies based on data sensitivity and user roles.

- **Context-Aware Access Control** further extends the capabilities of ABAC by incorporating real-time situational information into the access decision-making process. Whether it is restricting access based on geographical location, device security posture, or time of day, this model introduces an additional layer of security that adapts to the evolving threat landscape. Such contextual considerations are especially pertinent in scenarios where the risk associated with data access can vary significantly depending on external factors, offering a proactive approach to cybersecurity.

BALANCING ACCESS AND SECURITY

The move toward more dynamic access control models reflects a broader trend in cybersecurity: the need to balance protective measures with operational flexibility. In the context of knowledge generation, where insights are often derived from cross-referencing vast datasets, overly restrictive access controls can hinder the discovery process, potentially leading to incomplete or skewed analyses. ABAC and context-aware access control, with their emphasis on adaptable, context-sensitive permissions, provide a solution that safeguards data without impeding the analytical process.

- **Facilitating Comprehensive Analysis:** By allowing users access to a wider array of data within a secure framework, these advanced models enable more thorough and accurate knowledge generation. This is crucial in fields such as research, where access to diverse datasets can uncover groundbreaking insights.
- **Preventing Unauthorized Access:** At the same time, these models ensure that access is strictly aligned with user roles, objectives, and situational appropriateness, significantly reducing the risk of data breaches or leaks.

THE ROLE OF INHERITANCE IN ACCESS CONTROL

An underlying principle in the effective implementation of ABAC and context-aware access control across the DIKW pyramid is the concept of inheritance, albeit not in the traditional sense of direct rights transfer from one layer to another. Instead, inheritance here implies a logical progression of access principles that maintain consistency in security objectives while adapting to the specific needs of each layer.

- **Downward Inheritance:** While the specific access control model might shift from knowledge to data, the overarching principles – such as minimum necessary access and real-time risk assessment – inform security measures across all layers. This ensures a cohesive security posture that protects the organization's data assets while enabling the fluid exchange of information necessary for knowledge creation.

BROADER IMPLICATIONS OF ADVANCED ACCESS CONTROL MODELS

The adoption of ABAC and context-aware access control models signifies a paradigm shift in how organizations approach cybersecurity in the context of the DIKW pyramid. This evolution reflects a deeper understanding of the need for security measures that are as dynamic and complex as the data environments they protect. While these advanced models offer significant benefits, their implementation also introduces a set of challenges that organizations must navigate to realize their full potential.

- **Enhanced Security and Precision:** By considering a wide array of attributes and contextual factors, these models enable more precise access control decisions, significantly reducing the risk of unauthorized data access. This precision is particularly crucial in environments where data sensitivity varies widely and the potential impact of breaches is significant.
- **Flexibility and Scalability:** The adaptability of ABAC and context-aware models to various user roles, data types, and situational contexts makes them highly flexible and scalable. This adaptability ensures that access control systems can evolve alongside organizational growth and changes in the cybersecurity landscape.
- **Improved Compliance and Privacy Protection:** With their ability to dynamically enforce policies based on data sensitivity, user roles, and compliance requirements, these models support stronger compliance with regulatory standards. Moreover, they facilitate privacy protection by ensuring that access to personal or sensitive information is tightly controlled.

IMPLEMENTATION CHALLENGES

- **Complexity in Setup and Management:** The sophistication that makes ABAC and context-aware models so effective also contributes to their complexity. Designing and managing these systems requires a deep understanding of the organization's data architecture, user roles, and operational contexts, potentially increasing the burden on IT and security teams.
- **Need for Comprehensive Attribute and Policy Definition:** Implementing these models necessitates the comprehensive definition of attributes and access policies, a task that can be both time-consuming and prone to errors. Ensuring the accuracy and completeness of these definitions is critical for the system's effectiveness.
- **Balancing Security with Usability:** One of the perennial challenges in cybersecurity is balancing stringent security measures with user convenience and operational efficiency. Overly restrictive access controls can hinder productivity and innovation, particularly in knowledge-intensive environments where data exploration is key.

STRATEGIC CONSIDERATIONS FOR IMPLEMENTATION

To effectively leverage ABAC and context-aware access control models across the DIKW pyramid, organizations should consider the following strategic approaches:

- **Stakeholder Engagement:** Involve stakeholders from across the organization in the development and implementation of access control strategies. This inclusive approach ensures that the system aligns with diverse operational needs and user experiences.
- **Ongoing Training and Awareness:** Equip users and IT staff with the knowledge and skills to navigate the new access control environment. Regular training and awareness campaigns can help mitigate the risks associated with system complexity and ensure that users understand how to access the data they need securely.

- **Iterative Development and Continuous Improvement:** Adopt an iterative approach to developing and refining access control policies and configurations. Continuous monitoring and improvement can help organizations adjust to emerging threats, changing regulatory requirements, and evolving business objectives.

The broader implications of adopting ABAC and context-aware access control models underscore the potential for these advanced systems to transform organizational cybersecurity practices. By carefully navigating the implementation challenges and embracing strategic approaches to system design and management, organizations can harness the benefits of these models to secure their data environments effectively while promoting dynamic knowledge generation.

CONCLUSION: SAFEGUARDING THE JOURNEY THROUGH THE DIKW PYRAMID

The journey through the DIKW pyramid (Figure 21.1) represents more than just the transformation of raw data into actionable insights; it symbolizes the evolution of access control models in the face of burgeoning cybersecurity challenges. The shift from traditional frameworks like MAC and DAC to more dynamic, context-aware systems such as ABAC and context-aware access control underscores a fundamental change in how organizations approach the safeguarding of their digital assets (Figure 21.1). This chapter has explored the critical need for advanced access control models that align with the complexities of knowledge generation, ensuring that security measures facilitate rather than hinder the flow of information.

- **Embracing Complexity and Dynamism:** The progression toward ABAC and context-aware models reflects an acknowledgment of the complex interplay between security, privacy, and access. In an era where data-driven insights form the backbone

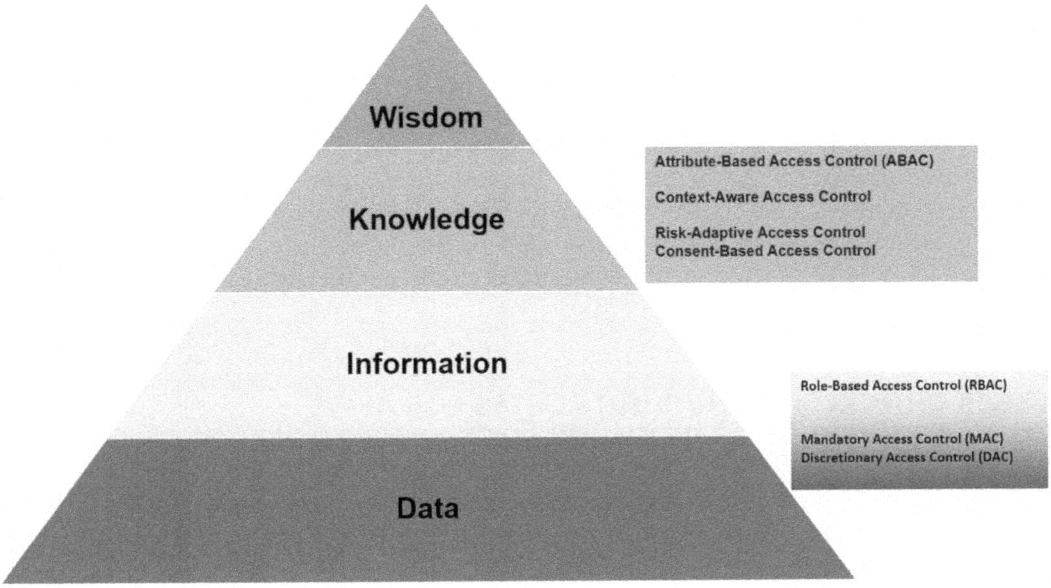

Figure 21.1 DIKW pyramid (data, information, knowledge, wisdom) in combination with different access models.

of decision-making, these models offer a nuanced approach to access control that balances the need for stringent security with the imperative for agility and exploration. Their implementation across the DIKW pyramid marks a strategic pivot toward systems that are as adaptive and multifaceted as the environments they protect.

- **Navigating Implementation Challenges:** While the benefits of these advanced access control models are manifold, their successful integration into organizational practices is not without challenges. The complexity of setup and management, the necessity for comprehensive attribute and policy definition, and the quest for balance between security and usability are hurdles that require thoughtful navigation. Overcoming these challenges demands a concerted effort that involves stakeholder engagement, ongoing training, and a commitment to iterative development and continuous improvement.
- **The Way Forward:** As organizations continue to ascend the DIKW pyramid, the role of sophisticated access control models in enabling secure, unrestricted knowledge generation becomes increasingly pivotal. The future of cybersecurity lies not in erecting barriers but in creating gateways – gateways that intelligently manage access based on context, role, and risk. This approach not only safeguards against threats but also empowers individuals and teams to harness the full potential of their data assets.
  - **A Unified Security Posture:** Achieving a coherent and aligned access control strategy across the DIKW layers is paramount. A unified security posture that seamlessly integrates advanced access control models ensures a consistent defense mechanism that spans the entirety of an organization's data ecosystem.
  - **Dynamic Response to Evolving Threats:** The adaptability of ABAC and context-aware models positions organizations to respond more effectively to the evolving threat landscape. By leveraging real-time data and analytics, these models offer a proactive stance on cybersecurity, one that anticipates and neutralizes threats even in their nascent stages.

IN CLOSING

The evolution of access control models within the context of the DIKW pyramid is a testament to the growing complexity of the digital domain and the innovative approaches being developed to navigate it. As we look to the future, the strategic implementation of ABAC and context-aware access control stands as a beacon for organizations striving to secure their digital journey. By embracing these models, organizations can unlock the door to advanced knowledge generation, ensuring that their voyage through the DIKW pyramid is both secure and enlightening.

CYBER CHUCKLE: RETHINKING ACCESS FOR AI – A GUIDE TO SURVIVING THE KNOWLEDGE JUNGLE

Welcome to the jungle, the knowledge jungle, where artificial intelligence (AI) beasts roam free, and the vines of data intertwine to form the elusive DIKW pyramid. It's a wild world out there, and as we embark on this safari through the dense foliage of access control, keep your machete sharp; we're cutting through complex concepts to ensure our AI overlords remain both powerful and tamed.

From the ground up: The DIKW safari

Our expedition begins at the base of the pyramid, wading through the Data swamps and Information rivers. Here, traditional access control mechanisms, like the alligators of the swamp, reign supreme. MAC and DAC, the guardians of the Data layer, ensure that only those with the right clearance can navigate these waters.

But as we trek further, ascending to the Knowledge canopy, the terrain shifts. The trees grow taller, the paths less defined. It's here we encounter the more adaptable creatures of the access control ecosystem – ABAC and context-aware models. These are the monkeys of the pyramid, swinging from branch to branch, ensuring that access is granted not by rigid hierarchy but by the situation at hand. They allow for a dynamic flow of information, enabling our AI entities to climb higher, reach further, and grasp the fruits of Wisdom at the pyramid's peak.

Navigating the knowledge layer: Tools for the trail

In this dense canopy, where traditional paths blur, how do we ensure that our AI guides don't lead us astray? Here are some tools for your trail:

- **The Compass of Goals:** Set clear destinations. Your AI needs to know whether it's aiming to help humanity or simply trying to win at Go.
- **The Map of Data:** Choose your data paths wisely. You don't want your AI learning from the swampy marshes of conspiracy theories.
- **The Machete of Transparency:** Keep your path clear. If your AI decides to take a detour, you should know why (and preferably before it ventures too deep into the conspiracy jungle).

The AI overlords: Keeping them happy (and secure)

As we make our way through this labyrinth of knowledge, it's essential to keep our AI overlords content. Here's some comical advice to ensure harmony between humanity and machine:

- **Feed Them Quality Data Snacks:** Just like humans, AI thrives on good nutrition. Feed it quality data, and you'll have a happy AI.
- **Give Them Clear Tasks to Play with:** An AI without a purpose is like a monkey without a forest. Give it clear, meaningful tasks and watch it swing gracefully through the data trees.
- **Respect Their Habitat:** The digital realm is their playground. Ensure it's safe, secure, and free from poachers looking to exploit AI vulnerabilities.

In this beautiful, bewildering world where machines learn and wisdom is digital, our journey through the DIKW pyramid teaches us the importance of adapting our strategies as we climb. As the lines between human and machine blur, remember: the goal is not to dominate but to coexist. To thrive in this ecosystem, we must learn not only to navigate but also to respect and understand the complex interplay of data, information, knowledge, and wisdom.

So, as we conclude our safari, let's remember to tread lightly, machete in hand, ready to pave new paths of understanding and coexistence in the AI jungle. Because with great AI power comes great meow-sponsibility.

Chapter 22

When the tide goes out

Dealing with the breach (with integrated mindfulness and human elements)

Spoiler Alert!

- **Ahoy, Mindful Mateys!** Dive into the aftermath of a cyber breach with a blend of calm mindfulness and a dash of humor.
- **Pause and Reflect:** Learn why taking a "breather" after a breach isn't just for yoga enthusiasts – it's for savvy cybersecurity pros too!
- **The Human Touch:** Find out how treating your team and stakeholders like humans (not robots) can actually help patch things up faster.
- **Stress Less, Protect More:** See how staying chill can make you a cybersecurity ninja, ready to tackle threats with a zen-like focus.

INTRODUCTION

Welcome aboard, cybersecurity professionals! As we navigate the complex and often turbulent waters of the cybersecurity sea, it's inevitable that sometimes the tide goes out, leaving behind the stark reality of exposed vulnerabilities and breaches in our digital defenses. This chapter delves into the aftermath of a cybersecurity breach, which can feel like navigating a ship through a storm with the guidance of stars obscured by clouds.

In the world of cybersecurity, breaches are the treacherous shoals and stormy seas that every organization hopes to avoid but must be prepared to face. Like seasoned sailors, cybersecurity professionals must not only have the skills to navigate these waters but also the wisdom to use every tool at their disposal to maintain course and resilience.

Section 1: Initial impact and the power of pause

When a breach hits, the initial impact can feel like a sudden storm upending a calm sea. Systems fail, data may be compromised, and the cybersecurity team is thrust into immediate action. But amidst this chaos, there is a powerful tool often overlooked: **mindfulness.**

Mindfulness in the midst of chaos

The first moments following the detection of a breach are critical. The usual response is a flurry of activity – securing systems, assessing damage, and mitigating any immediate threats. However, in this whirlwind of activity, the risk of knee-jerk reactions is high, and such reactions can sometimes exacerbate the problem.

DOI: 10.1201/9781003562184-29

To mitigate this, we introduce a strategic pause, a moment of mindfulness to anchor the team's response in clarity and precision:

1. **Immediate Structured Response**
   - **Scenario Description:** As soon as the breach is detected, alarms trigger, and the immediate protocol is enacted. The team's initial instinct is to jump straight into firefighting mode.
   - **Mindfulness Approach:** Implement a brief, structured pause. This 30-second mindfulness exercise involves deep breathing to center the team's focus and prevent panic-driven decisions.
   - **Implementation Steps**
     - **Breathe and Assess:** Engage in deep breathing exercises to calm the mind and sharpen focus.
     - **Prioritize Actions:** Use this moment of clarity to assess the situation based on verified alerts and prioritize actions based on the severity and impact of the breach.
2. **Mindfulness Integrated Response Plan**
   - **Calm Analysis:** With a clearer mind, the team analyzes the breach's scope. Mindfulness helps maintain this calm, ensuring that the analysis is methodical and thorough.
   - **Mindful Communication:** Communicate with stakeholders and the broader team in a composed manner, ensuring that information is disseminated clearly and without inducing unnecessary panic.

Mindfulness in practice

- **Practical Exercise:** Encourage the cybersecurity team to practice mindfulness regularly, not just in crisis. This could include starting meetings with a minute of silence or deep breathing exercises to foster a habit that will pay dividends in crisis.
- **Tool Integration:** Integrate mindfulness reminders into cybersecurity dashboards or incident response tools, prompting these pauses naturally within the workflow.

By incorporating mindfulness into the immediate response to a cybersecurity breach, organizations can ensure that their actions are as effective as possible, avoiding the pitfalls of hasty decisions made in panic. This approach doesn't just help in managing the technical aspects of a breach but also in maintaining the team's mental resilience amid high stress, setting the stage for a more structured and effective recovery process.

SECTION I: INITIAL IMPACT AND THE POWER OF PAUSE

In the turbulent aftermath of a cybersecurity breach, the seas are rough and the wind is strong. The crew is tense, and the ship's alarms are sounding. This is the moment when the initial impact of the breach hits full force, a moment that can define the course of subsequent actions and strategies. Here, the **Power of Pause** plays a pivotal role in steering the ship away from potential rocks and whirlpools that can exacerbate the situation.

Navigating through the storm with mindfulness

Mindfulness isn't just about being calm; it's about being present and aware, especially in the face of adversity. In the chaotic moments following a breach, mindfulness serves as an anchor, preventing the ship from being tossed aimlessly by the waves of panic and confusion.

1. **The Moment of Impact**
   - **Scenario Description:** The breach has been detected. Data is potentially being siphoned off by attackers, systems are compromised, and the usual business operations are in jeopardy. The cybersecurity team is on high alert, and there's an overwhelming urge to act swiftly to mitigate damages.
   - **Mindfulness Approach:** Before diving into reactive measures, the team takes a structured pause – a brief moment to breathe deeply and center themselves. This isn't inaction; it's a strategic pause that allows for a more measured and effective response.
   - **Implementation Steps**
     - **Deep Breathing:** Engage in a short session of deep breathing exercises to stabilize emotions and clear the mind.
     - **Focused Assessment:** Utilize this moment of clarity to perform a quick, focused assessment of the breach's immediate impact, prioritizing actions based on this calm analysis.
2. **Structured Response Using Mindfulness**
   - **Calm after the Initial Storm:** After the immediate pause, the response becomes more structured. The team methodically addresses the breach, with each step informed by a clear-headed assessment of priorities and risks.
   - **Balanced Reaction:** With a mindfulness approach, the reaction is balanced; it's neither a freeze nor a frantic scramble but a purposeful progression of steps designed to contain and assess the situation effectively.
   - **Tools and Techniques**
     - **Mindful Logging:** Maintain a log of actions taken, infused with mindfulness reminders to ensure that each step is deliberate and measured.
     - **Visualization Techniques:** Use visualization to mentally map out the breach scenario and potential responses, helping to anticipate challenges and plan responses effectively.

Incorporating mindfulness into the forensic investigation

Once the initial frenzy settles, the focus shifts to a meticulous forensic investigation, dissecting how the breach happened, the extent of the damage, and the vulnerabilities exploited. Mindfulness continues to play a crucial role here.

1. **Forensic Focus with Mindfulness**
   - **Detailed Investigation:** As the team delves into the forensic analysis, mindfulness helps maintain a high level of focus and prevents burnout, especially during extended investigations that can be mentally taxing.
   - **Mindful Breaks:** Incorporate mindful breaks to ensure that the investigative team remains sharp and doesn't overlook critical details due to fatigue.
   - **Implementation Steps**
     - **Scheduled Mindfulness Sessions:** During long hours of investigation, schedule brief mindfulness sessions to help the team reset and maintain attention to detail.

   - **Reflective Journaling:** Encourage the team to use reflective journaling to capture insights and reflections during the investigation, aiding in processing the information more effectively.
2. **Stress Management in High Pressure**
   - **Managing Stress:** The stress of dealing with a breach can affect decision-making and performance. Mindfulness helps manage this stress, ensuring that it doesn't lead to hasty conclusions or missed information.
   - **Techniques Employed**
     - **Guided Meditations:** Use short, guided meditations to help the team manage stress and return to the task with renewed focus.
     - **Physical Mindfulness Exercises:** Simple physical exercises, like stretching or mindful walking, can help alleviate the physical symptoms of stress, keeping the team both mentally and physically prepared to continue their work.

Communicating mindfully during crisis

Communication in the wake of a breach is as critical as the technical response. How the situation is communicated to stakeholders, staff, and possibly the public can significantly impact the organization's recovery and reputation.

1. **Transparent and Calm Communication**
   - **Clear Messaging:** Use mindfulness to ensure that communications are clear and free from the panic that can cloud important messages. This helps in maintaining transparency and trust with all stakeholders.
   - **Empathetic Engagement:** Recognize the human element; everyone is stressed, and mindful communication can help soothe anxieties and provide clear, compassionate updates on the situation.
   - **Implementation Steps**
     - **Mindful Messaging:** Before sending out communications, take a moment to review the message mindfully, ensuring it conveys the necessary information calmly and clearly.
     - **Active Listening Sessions:** Hold mindful listening sessions with stakeholders to address concerns and provide updates, ensuring that the dialogue is two-way and constructive.
2. **Feedback Loops and Mindful Reflection**
   - **Constructive Feedback:** Encourage a culture of constructive feedback, using mindfulness to process and reflect on the information received from all channels.
   - **Reflective Practice:** Incorporate reflective practices into team meetings to discuss what was handled well and what could be improved, promoting a cycle of continuous learning and improvement.

By integrating mindfulness into the initial response, forensic investigation, and communication strategies, the chapter outlines how organizations can navigate the choppy waters following a breach with greater calm, clarity, and effectiveness. This approach doesn't just repair the breach; it strengthens the team and the organization's resilience, preparing them to face future storms with confidence and strategic insight.

In this detailed journey through the initial impact and the power of pause, we see how a mindful approach can transform potential chaos into a structured path to recovery, proving that sometimes the best action is a well-considered pause.

SECTION 2: FORENSIC INVESTIGATION WITH A MINDFUL APPROACH

After navigating the initial shock of a cybersecurity breach and stabilizing the situation with a mindful pause, the next critical phase is the forensic investigation. This section delves into how integrating mindfulness into the forensic process can enhance focus, reduce stress, and lead to more accurate and insightful conclusions.

Detailed exploration with calm precision

Forensic analysis is akin to carefully charting the unexplored depths of the sea after a storm has passed. Each piece of data is a clue, and each clue could lead to understanding how the breach occurred, who was behind it, and how similar incidents can be prevented in the future.

1. Mindful Examination of Digital Traces
   - **Situation Overview:** The team begins the detailed work of tracing the steps of the attackers. This involves examining logs, reverse engineering malware, and piecing together the sequence of events.
   - **Mindfulness Approach:** Instead of rushing through this process, the team uses mindfulness to maintain a steady pace, ensuring that every piece of information is carefully analyzed without jumping to conclusions.
   - **Implementation Steps**
     - **Focused Analysis Sessions:** Conduct sessions where the team focuses on one aspect of the breach at a time, using mindfulness techniques like focused attention to prevent cognitive overload.
     - **Mindful Review:** After each session, take a short break to practice mindfulness, allowing the mind to rest and assimilate the information before moving on to the next piece of data.
2. Avoiding Bias and Ensuring Accuracy
   - **Challenge:** During forensic investigations, there's a risk of developing tunnel vision, where the team might focus too heavily on one lead or potential cause.
   - **Mindful Counteraction:** Mindfulness helps maintain an open mind, ensuring that investigators consider all possible angles and leads, reducing the risk of oversight or bias.
   - **Techniques Employed**
     - **Regular Mindful Check-ins:** At regular intervals, pause to conduct a brief mindful check-in to reassess assumptions and ensure that all potential leads are being objectively evaluated.
     - **Reflective Questioning:** Use mindfulness to encourage reflective questioning, asking "What else could this mean?" to explore alternative explanations and avoid premature conclusions.

Stress management and enhanced focus

The pressure to quickly resolve a cybersecurity breach can lead to stress and burnout among the team, particularly during the intensive forensic phase.

1. Managing Stress through Mindful Practices
   - **Situation Description:** The hours are long, the stakes are high, and the team is under significant pressure to deliver results swiftly.

- **Mindfulness Solution:** Incorporate stress-reducing mindfulness exercises to help the team maintain mental clarity and emotional stability.
- **Implementation Steps**
  - **Guided Breathing Exercises:** Integrate short guided breathing exercises to help reduce stress levels and refresh the mind, especially after analyzing complex or disturbing data.
  - **Mindful Movement Breaks:** Encourage brief, regular breaks for mindful walking or gentle stretching to reduce physical and mental tension.

2. **Enhanced Concentration and Detail Orientation**
   - **Challenge:** The minutiae of forensic data can be overwhelming, requiring high levels of concentration and attention to detail.
   - **Mindful Enhancement:** Use mindfulness to improve focus and prevent fatigue, ensuring that investigators stay alert and attentive to the intricate details of their work.
   - **Techniques Employed**
     - **Single-Tasking Focus Sessions:** Promote periods of single-tasking with mindfulness, where team members focus on one task at a time without multi-tasking distractions.
     - **Sensory Grounding Techniques:** Teach sensory grounding techniques to help team members stay present and focused, using simple prompts like feeling the texture of their chair or listening to the ambient sounds in the room.

Holistic communication and team support

Effective communication and team support are crucial in the forensic phase, where collaboration can significantly impact the speed and quality of the investigation.

1. **Mindful Communication among Team Members**
   - **Importance:** Clear and compassionate communication can enhance teamwork and lead to more comprehensive investigative outcomes.
   - **Mindful Approach:** Foster an environment where team members feel comfortable sharing insights and asking questions without judgment.
   - **Implementation Steps**
     - **Active Listening Sessions:** Hold active listening sessions where team members can share their findings and thoughts in a supportive, mindful environment.
     - **Empathy Exercises:** Practice empathy exercises to enhance mutual understanding and support among team members, recognizing the stress and challenges each member is facing.
2. **Supportive Feedback and Collaborative Analysis**
   - **Challenge:** Feedback is essential for refining the investigative process, but it can sometimes be a source of tension or defensiveness.
   - **Mindful Solution:** Approach feedback with mindfulness, ensuring it is constructive, balanced, and supportive.
   - **Techniques Employed**
     - **Feedback with Compassion:** Provide feedback in a way that is direct yet compassionate, helping team members learn and improve without feeling criticized or undervalued.
     - **Collaborative Problem-Solving:** Use mindfulness to foster a collaborative problem-solving atmosphere, where team members work together to unravel the complexities of the breach.

Reflective practice and learning from the breach

After the immediate forensic investigation is complete, mindfulness can guide the reflective practice, turning lessons learned into actionable insights for future prevention and resilience building.

1. **Reflective Analysis and Mindful Learning**
   - **Overview:** Once the primary forensic investigation is complete, it's time to reflect on what was learned and how these lessons can improve future security.
   - **Mindful Reflection:** Encourage a mindful review of the investigation, identifying key lessons and areas for improvement without rushing to move on or ignoring difficult truths.
   - **Implementation Steps**
     - **Post-Investigation Reflection Sessions:** Conduct sessions where the team can reflect on the investigation, using mindfulness to approach this process with openness and a willingness to learn.
     - **Mindful Debriefing:** Use debriefing techniques that include mindfulness exercises to help process the emotional and cognitive load of the investigation, ensuring that lessons are absorbed and integrated.
2. **Building Resilience for Future Challenges**
   - **Importance:** The end of one investigation is the beginning of building stronger defenses for the future.
   - **Mindful Approach:** Use the insights gained to build resilience, integrating mindfulness into the ongoing development of cybersecurity strategies.
   - **Techniques Employed**
     - **Resilience Workshops:** Hold workshops that use mindfulness to explore how resilience can be strengthened in the face of future breaches.
     - **Continuous Mindfulness Practice:** Encourage the ongoing practice of mindfulness within the team, ensuring that this approach becomes a foundational aspect of the cybersecurity culture.

By incorporating mindfulness into the forensic investigation and the broader response to a cybersecurity breach, organizations can enhance their ability to manage stress, maintain focus, communicate effectively, and learn from incidents. This mindful approach not only supports the technical aspects of breach response but also fosters a more resilient, reflective, and supportive cybersecurity team, ready to face future challenges with clarity and confidence.

SECTION 3: CRISIS COMMUNICATION – TRANSPARENCY AND RECOGNIZING HUMAN FACTORS

After stabilizing the initial impact of a cybersecurity breach and conducting a mindful forensic investigation, the focus shifts to crisis communication. This section explores how transparency, empathy, and the recognition of human factors are essential in effectively communicating with stakeholders during a cybersecurity crisis.

Transparent communication with stakeholders

Effective crisis communication is like navigating through foggy seas; transparency is the beacon that guides the way, ensuring that all stakeholders understand the situation and the steps being taken to address it.

1. **Openness and Honesty**
   - **Situation Overview:** In the aftermath of a breach, stakeholders, including customers, employees, and partners, are eager for information. They want to know what

happened, how it affects them, and what is being done to fix the issue and prevent future incidents.

- **Mindful Communication Approach:** Adopt a policy of openness and honesty, providing clear and concise information without overwhelming or causing unnecessary panic.
- **Implementation Steps**
  - **Regular Updates:** Provide regular updates through various channels (e.g., emails, press releases, social media) to keep stakeholders informed about the investigation's progress and recovery efforts.
  - **Clear, Jargon-Free Language:** Use clear, straightforward language to explain the situation, avoiding technical jargon that might confuse non-technical stakeholders.

2. **Balancing Transparency with Sensitivity**
   - **Challenge:** While transparency is crucial, there's also a need to balance openness with sensitivity to avoid spreading fear or revealing too much information that could exacerbate the situation.
   - **Mindful Balance:** Be transparent about the breach's impact while being mindful of stakeholders' concerns and fears.
   - **Techniques Employed**
     - **Empathetic Communication:** Tailor communications to address the emotional and informational needs of different stakeholder groups, showing understanding and empathy for their concerns.
     - **Controlled Disclosure:** Share essential details about the breach without disclosing sensitive information that could lead to further security vulnerabilities or panic.

Recognizing and addressing human factors

Acknowledging the human aspect of cybersecurity incidents helps in crafting communications that resonate with stakeholders, fostering a sense of shared responsibility and understanding.

1. **Humanizing the Incident**
   - **Situation Description:** Cybersecurity incidents often feel abstract and impersonal, leading to detachment or misunderstanding among stakeholders.
   - **Mindful Humanization:** Bring a human perspective to the incident, explaining how it affects individuals and what is being done to protect their data and privacy.
   - **Implementation Steps**
     - **Personal Stories:** Where appropriate, share anonymized stories or examples of how the breach has impacted individuals or the organization to illustrate the human side of cybersecurity.
     - **Employee Involvement:** Include messages from key team members involved in the breach response, sharing their commitment to resolving the issue and protecting stakeholders.
2. **Addressing Human Error and Responsibility**
   - **Challenge:** Human error is often a contributing factor in cybersecurity breaches, but addressing it can be sensitive and potentially demoralizing.
   - **Mindful Approach:** Discuss human error or vulnerabilities in a way that emphasizes learning and improvement, avoiding blame and fostering an environment of collective responsibility.
   - **Techniques Employed**
     - **Constructive Framing:** Frame any human errors as opportunities for learning and improvement, not as failures or reasons for punishment.

  - **Supportive Messaging:** Offer reassurance and support to those who might be feeling guilty or stressed due to their involvement in the incident, emphasizing the team's collective effort to learn and grow from the experience.

Building trust through effective communication

The ultimate goal of crisis communication is to build or restore trust with stakeholders by demonstrating competence, transparency, and empathy.

1. **Restoring Confidence**
   - **Objective:** Stakeholders need to feel confident that the organization can manage the breach and prevent future incidents.
   - **Mindful Confidence Building:** Use communications to demonstrate the organization's commitment to security and its proactive steps to enhance protective measures.
   - **Implementation Steps**
     - **Highlight Improvements:** Clearly outline any changes or improvements made to security protocols as a result of the breach.
     - **Future Plans:** Share plans for ongoing security assessments, trainings, and enhancements to reassure stakeholders about the organization's long-term commitment to cybersecurity.
2. **Maintaining a Dialogue**
   - **Importance:** Keeping an open line of communication with stakeholders is crucial for maintaining trust and managing the ongoing narrative around the incident.
   - **Mindful Dialogue:** Encourage questions and feedback from stakeholders, using this dialogue as an opportunity to further educate and engage them.
   - **Techniques Employed**
     - **Q&A Sessions:** Hold virtual Q&A sessions where stakeholders can ask questions and get real-time responses from the cybersecurity team.
     - **Feedback Loops:** Implement feedback mechanisms to gather stakeholders' thoughts and concerns, using this input to adjust recovery strategies and communication approaches.

Mindfulness in the midst of crisis

Integrating mindfulness into crisis communication helps maintain a calm, focused approach, ensuring messages are delivered thoughtfully and effectively.

1. **Mindful Preparation**
   - **Pre-Crisis Planning:** Develop mindfulness practices as part of crisis communication training, preparing the team to respond calmly and effectively under pressure.
   - **Techniques Employed**
     - **Scenario Planning:** Use mindfulness techniques during scenario planning to help team members envision various crisis communication challenges and develop balanced responses.
     - **Mindful Messaging:** Craft key messages in advance, using mindfulness to ensure they are thoughtful, empathetic, and aligned with the organization's values.
2. **During the Crisis**
   - **Active Mindfulness:** Apply mindfulness in real time to stay centered and present during crisis communications, avoiding reactive or overly emotional responses.

- **Implementation Steps**
  - **Breathing Exercises:** Encourage spokespersons and communicators to practice mindful breathing before delivering updates or interacting with stakeholders.
  - **Mindful Listening:** Practice mindful listening during stakeholder interactions to better understand their concerns and respond more effectively.

By recognizing human factors, fostering transparency, and employing a mindful approach to communication, organizations can navigate the turbulent waters of crisis communication with grace and effectiveness. This approach not only mitigates the immediate impacts of a cybersecurity breach but also strengthens relationships with stakeholders, laying the groundwork for a resilient and trusted cybersecurity posture.

The journey through crisis communication is a critical phase in responding to cybersecurity incidents. By embracing the principles outlined in this section, organizations can ensure that their response is not only effective but also compassionate and human-centered, turning the challenge of a breach into an opportunity for growth and improved stakeholder relations.

SECTION 4: LEGAL AND REGULATORY AFTERMATH WITH CALCULATED RISKS

In the wake of a cybersecurity breach, navigating the legal and regulatory aftermath requires a blend of precision, mindfulness, and strategic risk assessment. This section explores how organizations can handle legal obligations and regulatory requirements while maintaining a mindful approach to decision-making, particularly as risk perceptions evolve over the duration of a crisis.

Navigating legal challenges mindfully

After a cybersecurity incident, the legal ramifications can be daunting, with potential lawsuits, regulatory scrutiny, and compliance issues looming large.

1. **Legal Obligations and Compliance**
   - **Situation Overview:** Following a breach, organizations must adhere to a myriad of legal and regulatory obligations, from reporting the breach to relevant authorities to complying with data protection laws.
   - **Mindful Legal Navigation:** Approach these obligations with a clear, focused mindset. Being mindful of legal responsibilities helps in prioritizing actions and communicating transparently with regulatory bodies.
   - **Implementation Steps**
     - **Immediate Reporting:** Ensure timely reporting to regulatory bodies as required by law (e.g., General Data Protection Regulation (GDPR), Health Insurance Portability and Accountability Act of 1996 (HIPAA)). Use mindfulness to maintain accuracy and completeness in these reports, avoiding rushed submissions.
     - **Documentation and Evidence Collection:** Meticulously document the breach investigation process and outcomes. Mindful attention to detail here can prevent discrepancies and provide solid evidence in legal proceedings.
2. **Handling Lawsuits and Legal Proceedings**
   - **Challenge:** Potential lawsuits from affected parties or stakeholders can exacerbate the post-breach scenario.

- **Strategic and Mindful Approach:** Engage with legal counsel mindfully, focusing on building a robust defense by presenting clear, well-documented evidence of the breach response efforts.
- **Techniques Employed**
  - **Legal Strategy Sessions:** Conduct regular strategy sessions with legal teams, employing mindfulness to stay focused and avoid being overwhelmed by the complexities of legal arguments.
  - **Empathetic Stakeholder Engagement:** When communicating with parties involved in lawsuits, use a mindful, empathetic approach to address their concerns and negotiate resolutions.

Risk decision dynamics: Navigating risk fatigue

The perception of risk can change over the course of managing the breach's aftermath, initially being risk-averse and gradually moving toward risk acceptance.

1. **Initial Risk Aversion**
   - **Description:** In the immediate aftermath of a breach, organizations are typically highly risk-averse, prioritizing secure and conservative measures to mitigate any further damage.
   - **Mindful Risk Assessment:** Use mindfulness to maintain a balanced perspective, ensuring that initial risk aversion doesn't lead to overly cautious or paralyzing strategies.
   - **Implementation Steps**
     - **Comprehensive Risk Analysis:** Conduct thorough risk assessments with a calm, focused approach to identify and mitigate potential threats without rushing into restrictive measures.
     - **Scenario Planning:** Engage in mindful scenario planning to anticipate various outcomes and prepare measured responses.
2. **Evolving Risk Willingness**
   - **Challenge:** As the crisis continues, there's a natural tendency to become desensitized to risk, potentially leading to riskier decisions as time progresses.
   - **Balanced Risk Management:** Employ mindfulness to recognize this shift in risk perception and maintain a balanced approach to decision-making.
   - **Techniques Employed**
     - **Regular Risk Review Meetings:** Hold periodic risk assessment reviews, using mindfulness to ensure these sessions are focused and productive, helping to recalibrate risk tolerance levels.
     - **Feedback Loops:** Establish open feedback mechanisms where team members can express their concerns or observations about evolving risk attitudes.

Strategic use of mindfulness in risk management

Incorporating mindfulness into risk management not only helps in maintaining focus and balance but also aids in adapting to the changing landscape of legal and regulatory challenges.

1. **Mindful Legal and Risk Strategies**
   - **Approach:** Develop legal and risk management strategies that are infused with mindfulness, ensuring that each decision is considered and balanced.

- **Implementation Steps**
  - **Mindful Legal Review:** Before making any legal decisions or taking actions, engage in a brief period of mindfulness to center thoughts and ensure a clear, unbiased approach.
  - **Risk Management Workshops:** Organize workshops where mindfulness techniques are taught alongside risk management skills, enhancing the team's ability to handle stress and uncertainty.

2. **Adaptive Legal Compliance**
   - **Situation:** Regulatory environments are often dynamic, with rules and guidelines evolving in response to emerging threats and new technologies.
   - **Mindful Adaptation:** Stay mindful of these changes and adapt compliance strategies in a thoughtful, measured manner.
   - **Techniques Employed**
     - **Regulatory Monitoring:** Use a mindful approach to monitor regulatory updates, ensuring that the organization remains compliant without hastily overhauling established protocols.
     - **Stakeholder Communication:** Communicate changes in compliance and legal strategies to all stakeholders in a mindful, clear manner, ensuring they understand the reasons and implications.

Preparing for future legal and regulatory challenges

The legal and regulatory journey following a cybersecurity breach is not just about addressing immediate concerns but also about preparing for future challenges.

1. **Long-Term Legal Strategy**
   - **Goal:** Develop a long-term legal strategy that anticipates future regulatory changes and potential legal challenges.
   - **Mindful Long-Term Planning:** Incorporate mindfulness into strategic planning sessions to ensure that long-term strategies are thoughtful and proactive.
   - **Implementation Steps**
     - **Legal Trend Analysis:** Conduct regular reviews of legal trends and potential regulatory changes, using mindfulness to stay focused and extract relevant insights.
     - **Collaborative Legal Planning:** Work collaboratively with legal, compliance, and cybersecurity teams to develop integrated strategies that reflect a balanced, mindful approach to future challenges.
2. **Building a Culture of Mindful Legal Compliance**
   - **Objective:** Foster a culture where legal compliance is approached mindfully, ensuring that it's integrated into daily operations and decision-making processes.
   - **Mindful Culture Development**
     - **Training and Workshops:** Offer training sessions and workshops on mindful legal compliance, helping employees understand the importance of mindfulness in maintaining compliance and managing risks.
     - **Mindful Decision-Making Processes:** Embed mindfulness practices into decision-making processes, particularly those involving legal and regulatory assessments, to encourage thoughtful and thorough analysis.

By embracing a mindful approach to the legal and regulatory aftermath of a cybersecurity breach, organizations can navigate this complex landscape with greater clarity and balance. This approach helps mitigate risk, ensures compliance, and prepares the organization for future challenges, all while maintaining a focus on the human elements that underpin effective crisis management.

SECTION 5: THE HUMAN COST – MINDING YOUR PEOPLE

In the turbulent aftermath of a cybersecurity breach, the focus often zooms in on technical recovery and legal maneuvers. Yet, an equally critical aspect is the human cost – the impact on the people within the organization who face stress, anxiety, and often a sense of personal responsibility for the breach. This section delves into how integrating mindfulness and understanding human elements can significantly aid in healing and strengthening the team post-breach.

Understanding the emotional impact

A cybersecurity breach doesn't just disrupt systems; it disrupts lives. Employees may feel violated, scared, or guilty, especially if human error contributed to the breach.

1. **Initial Emotional Responses**
    - **Description:** In the immediate aftermath, there's a whirlwind of emotions. Staff may experience shock, confusion, and distress.
    - **Mindful Acknowledgment:** Encourage a culture where these feelings are acknowledged and not dismissed. Mindfulness can help individuals recognize and accept their emotions without judgment.
    - **Implementation Steps**
        - **Open Forums:** Create safe spaces like open forums or town-hall meetings where employees can express their feelings and concerns.
        - **Active Listening Sessions:** Conduct mindful listening sessions where management actively listens to employee concerns, helping to validate their feelings and foster a supportive environment.
2. **Stress and Anxiety Management**
    - **Challenge:** Prolonged stress and anxiety can lead to burnout and reduced productivity, impacting the overall recovery process.
    - **Mindful Stress Reduction:** Introduce mindfulness exercises that can help reduce stress and anxiety levels. Practices like guided meditations or mindful breathing exercises can be particularly beneficial.
    - **Techniques Employed**
        - **Regular Mindfulness Breaks:** Implement short, guided mindfulness sessions throughout the workday to help staff manage stress.
        - **Mindfulness Training Workshops:** Offer workshops on mindfulness techniques tailored to stress reduction, enhancing employees' ability to cope with post-breach pressures.

Supporting employee well-being

The path to recovery is not just about restoring systems but also about supporting the people who operate them. Mindfulness can play a key role in this holistic recovery approach.

1. **Employee Support Programs**
    - **Description:** Comprehensive employee support programs are essential for helping staff navigate the personal and professional challenges posed by the breach.

- **Mindful Support Initiatives:** Ensure these programs include mindfulness-based approaches to provide a more rounded support system.
- **Implementation Steps**
  - **Counseling Services:** Provide access to counseling services that include mindfulness-based stress reduction techniques.
  - **Peer Support Groups:** Facilitate the formation of peer support groups where employees can share experiences and coping strategies in a mindful, supportive setting.

2. **Maintaining Productivity and Focus**
   - **Challenge:** Maintaining productivity and focus can be difficult when employees are dealing with the fallout from a breach.
   - **Mindful Work Practices:** Implement mindful working practices to help employees maintain concentration and performance.
   - **Techniques Employed**
     - **Mindful Task Management:** Encourage the use of mindfulness techniques in managing tasks, such as focusing on one task at a time and taking mindful breaks to reset.
     - **Environment Optimization:** Create a work environment that supports mindfulness, with quiet spaces for meditation or reflection, enhancing the overall work atmosphere.

Building resilience and reducing guilt

The breach can often leave employees feeling guilty, especially if their actions were directly involved. Building resilience and reducing guilt are key to moving forward.

1. **Addressing Guilt and Responsibility**
   - **Description:** It's common for employees to feel guilty after a breach, particularly if their mistake was a contributing factor.
   - **Mindful Approach to Guilt:** Use mindfulness to help employees process and move beyond guilt, focusing on learning and growth rather than blame.
   - **Implementation Steps**
     - **Mindful Reflection Sessions:** Organize sessions where employees can reflect on the incident in a structured, mindful manner, helping them learn from the experience without dwelling on guilt.
     - **Empathy Training:** Provide training on empathy and self-compassion, emphasizing that mistakes are part of the human experience and can lead to valuable growth.
2. **Enhancing Organizational Resilience**
   - **Goal:** Build a resilient organization where employees are equipped to handle future challenges without fear or hesitation.
   - **Mindful Resilience Building:** Incorporate mindfulness into resilience training, ensuring that employees are prepared for future incidents in a balanced and confident manner.
   - **Techniques Employed**
     - **Resilience Workshops:** Conduct workshops that combine resilience training with mindfulness practices, helping employees develop a toolkit for coping with and bouncing back from future incidents.
     - **Regular Resilience Check-Ins:** Establish a routine of regular check-ins where employees can discuss their resilience journey, facilitated by mindfulness exercises to enhance this process.

To sum up

Integrating mindfulness into the approach to managing the human cost of a cybersecurity breach offers a path to recovery that is compassionate, holistic, and effective. By focusing on the emotional well-being of employees, supporting their mental health, and building a resilient workforce, organizations can emerge from the aftermath of a breach stronger and more unified.

In the next section, we will explore how these principles are woven into lessons learned and the building of future resilience, ensuring that the organization not only recovers but thrives in the face of ongoing cybersecurity challenges.

SECTION 6: LESSONS LEARNED AND BUILDING RESILIENCE WITH MINDFULNESS

In the aftermath of a cybersecurity breach, the journey from recovery to resilience is pivotal. This section explores how integrating lessons learned with mindfulness can fortify an organization's defenses and enhance its ability to navigate future challenges with greater agility and awareness.

Extracting lessons from the breach

Post-breach analysis is crucial for understanding what went wrong and how similar incidents can be prevented in the future. Mindfulness can play a significant role in this reflective process, ensuring that lessons are learned in a thoughtful, constructive manner.

1. **Comprehensive Debriefs**
    - **Description:** After a breach, conducting comprehensive debrief sessions can help distill key lessons and facilitate a shared understanding of the incident.
    - **Mindful Reflection:** Use mindfulness to guide these debriefs, promoting a non-judgmental atmosphere where all team members feel safe to share their perspectives and insights.
    - **Implementation Steps**
        - **Structured Debrief Sessions:** Organize structured debrief sessions that encourage mindful listening and reflection, allowing team members to discuss the incident openly and without fear of blame.
        - **Documentation of Lessons:** Ensure that the lessons learned are documented in a clear, accessible format and include mindful reflections on how emotions and reactions influenced the breach response.
2. **Identifying and Addressing Root Causes**
    - **Challenge:** Identifying the root causes of a breach is essential for preventing future incidents.
    - **Mindful Analysis:** Employ mindfulness to approach this analysis with an open, curious mind, free from preconceptions or biases, allowing for a clearer understanding of underlying issues.
    - **Techniques Employed**
        - **Root Cause Analysis Workshops:** Conduct workshops where mindfulness techniques are used to help participants focus deeply on analyzing the data without jumping to conclusions.
        - **Mindful Brainstorming:** Facilitate mindful brainstorming sessions to explore potential root causes and solutions, ensuring a comprehensive and inclusive approach to problem-solving.

Building resilience through mindfulness

With lessons learned, the focus shifts to building resilience – enhancing the organization's capacity to adapt to and recover from future cybersecurity challenges. Mindfulness is integral to developing this resilience, providing the mental and emotional groundwork for robust recovery strategies.

1. **Developing a Resilient Mindset**
   - **Description:** A resilient mindset is crucial for navigating the uncertainties of the cybersecurity landscape.
   - **Mindful Resilience Training:** Incorporate mindfulness into resilience training programs to help employees develop a mindset that embraces challenges as opportunities for growth.
   - **Implementation Steps**
     - **Resilience and Mindfulness Workshops:** Offer workshops that combine resilience-building exercises with mindfulness practices, enhancing the ability to respond to stress and change with flexibility and strength.
     - **Mindful Mentoring Programs:** Establish mentoring programs where experienced staff use mindfulness techniques to guide less experienced team members in developing a resilient approach to their work.
2. **Reinforcing Emotional and Psychological Strength**
   - **Goal:** Strengthen the emotional and psychological resilience of the workforce to better cope with future incidents.
   - **Mindful Emotional Support:** Use mindfulness to provide ongoing emotional support to employees, helping them manage stress and recover from the psychological impacts of the breach.
   - **Techniques Employed**
     - **Regular Mindfulness Sessions:** Implement regular mindfulness sessions focused on emotional resilience, teaching employees how to maintain equilibrium in the face of cybersecurity threats.
     - **Support Networks:** Create support networks that use mindfulness-based approaches to provide emotional support, helping employees feel connected and supported within the organization.

Integrating mindfulness into organizational culture

To truly build resilience, mindfulness should be woven into the fabric of the organization's culture, making it a foundational aspect of how the organization operates and responds to challenges.

1. **Cultivating a Mindful Culture**
   - **Description:** A mindful organizational culture enhances collective resilience and the capacity to deal with cyber threats effectively.
   - **Mindful Culture Initiatives:** Implement initiatives that promote mindfulness throughout the organization, from the C-suite to entry-level positions.
   - **Implementation Steps**
     - **Mindfulness Training for All Levels:** Provide mindfulness training across all levels of the organization to foster a universally resilient mindset.
     - **Mindful Leadership Programs:** Encourage leaders to practice mindfulness and lead by example, promoting a culture where mindfulness is valued as a key component of strategic decision-making.
2. **Ongoing Mindfulness and Resilience Education**
   - **Goal:** Ensure that mindfulness and resilience are not just temporary focuses but enduring values.

- **Continued Learning and Practice:** Facilitate ongoing education and practice in mindfulness and resilience to maintain these as core organizational strengths.
- **Techniques Employed**
  - **E-Learning Modules:** Develop e-learning modules on mindfulness and resilience, allowing employees to learn and practice at their own pace.
  - **Mindfulness Retreats:** Organize annual mindfulness retreats to deepen employees' practice and understanding of mindfulness as a key tool in cybersecurity resilience.

To sum up

By integrating mindfulness into the process of learning from breaches and building resilience, organizations can create a more robust framework for navigating the cybersecurity challenges of the future. This approach not only enhances the technical aspects of cybersecurity but also strengthens the human elements, preparing the entire organization to face future threats with confidence and clarity.

CYBER CHUCKLES

To wrap up this chapter on a lighter note, here are some humorous takes on the key concepts we've discussed. These cyber chuckles reinforce the importance of mindfulness and a human-centered approach in cybersecurity, ensuring you leave with a smile and some food for thought!

- **Mindful Breaching:** "Why did the mindful hacker refuse to rush the breach? Because they knew the importance of taking a 'paws' to reflect before acting! And yes, they double-checked if their malware also felt ready."
- **Risk Decisions over Time:** "How does a mindful cybersecurity team change its risk decisions over time? Initially, they're as cautious as a cat in a new home. By the end, they're leaping like a lion, ready to take on the savannah of cyberspace!"
- **Humanizing the Attacker:** "Why did the cybersecurity team send a 'Get Well Soon' card to the attacker? Because after a mindful reflection, they realized that behind every breach is just another person who forgot to drink their calming tea."
- **Power of Pause:** "Why did the cybersecurity team start practicing yoga after a breach? Because they found out the power of 'pausing' can help stretch their patience and not just their security protocols! Now, their firewall does yoga too, bending but not breaking under pressure."
- **Integration with Mindfulness:** "What did the cybersecurity professional say after meditating on a complex breach? 'To understand the root cause, sometimes you must first understand the root chakra.'"

These lighthearted jokes encapsulate the essence of integrating mindfulness into cybersecurity practices. By maintaining a balanced perspective and embracing the human elements, cybersecurity professionals can navigate the aftermath of breaches more effectively and with a bit of humor to boost morale!

Now, with these principles and a smile, you're better equipped to face any cybersecurity challenges that come your way. Remember, in the world of cybersecurity, staying mindful and human-centered isn't just wise; it's also a way to keep the journey interesting and engaging. Happy navigating!

Chapter 23

Navigating cybersecurity

The Integrated Cybersecurity Compass (ICC)

Spoiler Alert!

- Navigate the cybersecurity ocean with more than just a paddle; grab your ICC and set sail on strategic seas!
- Warning: Nautical puns and metaphors ahead (it's smooth sailing from here, promise!).
- Mastering the ICC is like being the captain of your own cyber ship – knowing when to batten down the hatches and when to let the sales fly.
- Learn how to steer through storms of data breaches and pirate hackers with tools like the Cynefin Framework and Resilience Engineering.
- Ready to become the master and commander of your cybersecurity destiny? Let's embark on this voyage together!

INTRODUCTION

Welcome aboard the voyage through the turbulent waters of cybersecurity, where the digital sea is vast and the threats, like unpredictable storms, are ever-present and evolving. In this dynamic environment, traditional security measures – our digital lighthouses and buoys – play crucial roles but often struggle to adapt swiftly to the changing tides of cyber threats. The need for a more sophisticated navigation tool is clear, a tool that not only guides but adapts, anticipates, and acts with precision. Enter the Integrated Cybersecurity Compass (ICC), not merely a compass but a full-fledged nautical chart for the cybersecurity realm.

The ICC is a comprehensive framework designed to help organizations navigate the complex, often murky waters of cybersecurity. It draws on established navigational aids such as the Cynefin Framework and principles of Resilience Engineering and integrates them with cutting-edge economic theories and core security principles. This chapter will unfold how the ICC, much like a seasoned captain scanning the horizon with a sextant, enables businesses to chart a proactive and informed course through the cyber threats that loom in the digital deep.

In the realm of cybersecurity, the weather patterns are akin to the types of threats organizations face – some as overt and destructive as hurricanes, others as sudden and disorienting as squalls. The ICC uses the Cynefin Framework to categorize these threats, helping to tailor strategies that are as variable and complex as the threats themselves. Meanwhile, the currents and undercurrents, those unseen forces that can alter a ship's course without warning, are mirrored in the interdependencies within an organization's systems. Resilience Engineering helps map these hidden reefs and currents, ensuring that changes in security measures enhance safety rather than inadvertently creating new vulnerabilities.

DOI: 10.1201/9781003562184-30

Moreover, course correction is an ongoing necessity in both nautical and cybersecurity endeavors. The ICC advocates for an adaptive response design, utilizing real-time monitoring and the Simplified Complex Adaptive Systems (CAS) Model, especially for navigating the complex and chaotic threats identified by the Cynefin Framework. This approach ensures that as new threats emerge or old threats evolve, the security measures in place can be dynamically adjusted – much like a ship's rudder steering through choppy waters.

By integrating these diverse but complementary components, the ICC equips organizations not only with the tools to respond to immediate threats but also with the foresight to anticipate and prepare for potential future challenges. The journey through cybersecurity is not a sprint but a marathon – a continuous voyage where vigilance, adaptability, and a deep understanding of the landscape are key to maintaining course and ensuring safe passage through the digital age.

As we set sail into the details of the ICC in the following sections, we invite you to adjust your mental compasses and prepare for a deep dive into a strategy that transcends traditional cybersecurity measures, offering a holistic, adaptable, and forward-thinking approach to navigating the ever-expanding cyber seas.

SECTION I: NAVIGATIONAL TOOLS OF THE ICC – THE CYNEFIN FRAMEWORK AND RESILIENCE ENGINEERING

As the first mate of our navigational toolkit, the Cynefin Framework stands at the helm, guiding us through the complex waters of cybersecurity threats. Originating from the Welsh word for "habitat," Cynefin offers a rich, context-based approach to decision-making, illuminating the landscape in which cybersecurity professionals operate. It's an essential part of the ICC, helping to categorize the myriad threats and challenges into distinct domains, each requiring a tailored strategic response.

Understanding the domains of Cynefin

The Cynefin Framework divides issues into five domains, each representing different states of system behavior: Simple, Complicated, Complex, Chaotic, and Disorder. In the context of cybersecurity, these domains help professionals assess situations and apply appropriate strategies effectively.

1. **Simple (Obvious) Domain:** This is the realm of best practices where cause and effect are well understood and clearly visible. Actions to be taken are straightforward and predictable, much like following a well-marked chart through calm waters. In cybersecurity, simple threats are those with known solutions, such as standard software patching or enforcing strong password policies.
2. **Complicated Domain:** Here, expert navigation is required. Though the relationship between cause and effect is not immediately apparent, it can be discerned through analysis or expertise. This domain is akin to navigating by celestial navigation, requiring theoretical knowledge and practical experience. Cybersecurity examples include intricate system configurations or network designs where expert diagnostic skills are essential to identify vulnerabilities.
3. **Complex Domain:** In this domain, cause and effect can only be perceived in retrospect, not in advance. Like navigating through a foggy sea without a clear path, handling complex cybersecurity issues requires probing safely, sensing responses, and

then responding. Techniques such as adaptive security architectures, which learn and evolve based on new data, are effective in this unpredictable environment.

4. **Chaotic Domain:** Amidst the turmoil where no clear cause and effect relationships exist, rapid, decisive action is required to establish order. This is similar to steering a ship through a storm, where immediate, often innovative actions are necessary to prevent disaster. In cybersecurity, this might involve responding to a zero-day exploit or mitigating a ransomware attack in progress.
5. **Disorder:** The central domain, where it is unclear which of the other four contexts apply. Here, the first task is to break down the situation to identify which domain you are truly in. It's the nautical equivalent of finding yourself in uncharted waters, necessitating a careful assessment to determine your position and the best course of action.

Applying the Cynefin Framework in cybersecurity

Integrating the Cynefin Framework into the ICC empowers organizations to classify their cybersecurity challenges accurately and apply the most effective strategies. By recognizing the nature of the threat landscape – whether simple, complicated, complex, or chaotic – security teams can choose the appropriate tools and responses, much like a captain selects the right sails and navigation tools based on weather conditions.

- **For Simple and Complicated situations**, established protocols and expert analyses lead the way.
- **In Complex scenarios**, continuous monitoring and adaptive strategies are paramount.
- **Chaotic situations** require immediate and decisive action to stabilize the environment and move the situation into a domain where it is more manageable.

Through the lens of the Cynefin Framework, cybersecurity is not just a technical challenge but a strategic voyage requiring nuanced understanding and flexible tactics. This model does not provide a one-size-fits-all map but rather teaches us how to chart the course dynamically based on the nature of the seas we navigate. As we delve deeper into the ICC, we will see how other tools and strategies integrate with the insights provided by Cynefin to enhance our cybersecurity navigation systems.

Resilience Engineering in the Integrated Cybersecurity Compass

Following the guiding lights provided by the Cynefin Framework, we delve deeper into the navigational toolbox of the ICC by harnessing the principles of Resilience Engineering. This approach shifts our focus from merely reacting to cybersecurity incidents to proactively designing and maintaining systems that anticipate, adapt to, and withstand disruptions.

Foundations of Resilience Engineering

Resilience Engineering is rooted in the capability of a system not only to handle expected risks but also to adapt to unexpected threats in real time. It emphasizes the importance of system behavior under varying conditions and the necessity of bouncing back from cyber-attacks swiftly and effectively. Just as a ship's hull is designed not just to float but also to withstand the roughest of seas, cybersecurity systems require a design that accommodates unexpected and dynamic cyber threats.

Key components of Resilience Engineering in cybersecurity

1. **Understanding System Interdependencies:** The first step in Resilience Engineering involves a thorough analysis of how different components of an organization's IT systems interact. This process identifies how these interactions could potentially lead to vulnerabilities. For instance, the failure of a single authentication server could affect multiple services that rely on it for user verification, much like how a malfunction in a ship's engine affects its navigational capabilities and overall seaworthiness.
2. **Identifying Potential Failure Points:** This component focuses on pinpointing critical elements within the system architecture that, if compromised, could lead to significant disruptions. By understanding these points, cybersecurity teams can implement targeted safeguards and develop contingency plans that enhance system resilience – similar to a ship having watertight compartments to prevent sinking if the hull is breached.
3. **Adaptive Response and Recovery:** Leveraging insights from the Cynefin Framework, Resilience Engineering promotes an adaptive approach to cybersecurity. This means designing security measures that are not static but evolve based on the threat landscape and real-time system performance. Adaptive response mechanisms ensure that cybersecurity defenses can adjust quickly to new or escalating threats, much like a ship adjusting its sails to cope with changing wind conditions.

Implementing Resilience Engineering

To effectively implement Resilience Engineering within the ICC framework, organizations need to take several practical steps:

- **Routine Stress Testing and Scenario Analysis:** Just as naval architects test ship designs under various conditions, cybersecurity teams must simulate attack scenarios to understand how their systems would perform under stress. This testing helps identify weak points in the security posture that need reinforcement.
- **Integration of Real-Time Monitoring Tools:** Continuous monitoring of network and system activities allows for the early detection of anomalies that could indicate a cybersecurity threat. Real-time data feeds into adaptive response systems, enabling quick adjustments to defensive measures.
- **Fostering a Recovery Culture:** Building resilience is not just about preventing breaches but also about minimizing their impact and recovering quickly when they occur. This involves having robust disaster recovery and business continuity plans that are regularly updated and tested.
- **Training and Awareness:** Human elements play a crucial role in the resilience of cybersecurity systems. Regular training sessions for IT staff and general employees help raise awareness about potential cyber threats and the importance of adhering to security policies.

Incorporating Resilience Engineering into the ICC equips organizations with a proactive and robust methodology to address the dynamic challenges of cybersecurity. By understanding interdependencies, identifying failure points, and implementing adaptive responses, the resilience of cybersecurity systems is significantly enhanced. This approach ensures that organizations are not only prepared to navigate through the calm waters of known threats but are also capable of withstanding the unpredictable storms of the digital world.

SECTION 2: ECONOMIC AND BEHAVIORAL CONSIDERATIONS

Game theory and cybersecurity

As we set sail into deeper waters within the ICC, we encounter the complex interplay of economic and behavioral factors that significantly influence cybersecurity strategies. Game theory, a critical component of this exploration, offers a rigorous framework for understanding and navigating the strategic decisions made by both attackers and defenders in the cybersecurity domain.

Game theory: The strategic battlefield

Game theory explores the decision-making processes in situations where the outcome for each participant depends on the actions of others. In the context of cybersecurity, it provides a structured method to predict potential behaviors of cyber attackers and to devise effective defense mechanisms.

1. **Understanding the Adversary:** Cybersecurity is not just a technological challenge but also a strategic one, where understanding the adversary's motivations and potential actions is crucial. Game theory helps in modeling these interactions as a series of strategic games, where both attackers and defenders choose their actions based on the expected moves of the other.
2. **Nash Equilibrium in Cybersecurity:** This concept from game theory identifies the optimal strategy for each player, assuming the other players are also playing their optimal strategies. In cybersecurity, achieving a Nash Equilibrium means finding a state where neither the attacker nor the defender can unilaterally deviate from their chosen strategy without making themselves worse off. For instance, a well-implemented security measure that deters potential attacks might represent such an equilibrium.
3. **Zero-Sum vs. Non-Zero-Sum Scenarios:** Cybersecurity scenarios can be modeled as zero-sum games, where the gain of one player is exactly the loss of the other, or non-zero-sum games, where cooperative strategies might lead to better outcomes for all involved parties. Understanding these dynamics can guide the allocation of resources and the development of collaborative strategies among different entities.

Implementing game theory in cybersecurity strategies

To effectively incorporate game theory into cybersecurity practices, organizations should consider the following approaches:

1. **Scenario Planning and Simulations:** By simulating different attack and defense scenarios, organizations can anticipate potential adversarial strategies and develop more robust defenses. Game theory provides the theoretical framework for these simulations, helping to identify the most effective strategies in various contexts.
2. **Resource Allocation:** Game theory can guide the strategic allocation of cybersecurity resources. By understanding the potential moves of adversaries, organizations can prioritize investments in security measures that offer the highest payoff in terms of risk reduction.
3. **Collaborative Defense Strategies:** Game theory encourages the exploration of cooperative strategies among multiple stakeholders. In cybersecurity, this can translate into shared threat intelligence networks and collaborative response strategies, optimizing the collective defense posture.

Behavioral economics: Human factors in cybersecurity

Beyond the strategic interactions modeled by game theory, cybersecurity is profoundly influenced by human behavior. Behavioral economics provides insights into how cognitive biases and decision-making errors can affect security practices.

1. **Cognitive Biases**: Humans are not always rational actors, and their decisions can be swayed by biases such as overconfidence, anchoring, or aversion to loss. In cybersecurity, these biases can lead to underestimating risks, overlooking potential threats, or failing to adopt necessary security measures.
2. **Designing User-Centric Security**: By understanding these human factors, cybersecurity professionals can design systems that are not only secure but also user-friendly, reducing the likelihood of user error and increasing the overall effectiveness of security protocols.
3. **Incentive Structures**: Behavioral economics also studies how different incentive structures can influence behavior. In cybersecurity, carefully designed incentives can encourage better security practices, such as rewarding employees for identifying security flaws or adhering to best practices.

Conclusion: A comprehensive economic and behavioral framework

The integration of game theory and behavioral economics into the cybersecurity strategy provides a comprehensive framework that accounts for the complex interactions and human elements inherent in security challenges. By understanding and anticipating the strategic moves of adversaries and addressing the human factors that influence security practices, organizations can develop more effective and resilient cybersecurity measures. This approach not only enhances the technical defenses but also aligns them with the human behaviors and strategic decisions that underpin the entire cybersecurity landscape.

Cost-benefit analysis in cybersecurity investment

Venturing further into the economic and behavioral waters of cybersecurity, we engage with the pragmatic aspect of cybersecurity investments through cost-benefit analysis (CBA). This analytical tool is pivotal for organizations to rationalize expenditures in cybersecurity by evaluating the potential return on investment (ROI) and minimizing financial risks associated with cyber threats.

Understanding cost-benefit analysis

CBA in cybersecurity involves a systematic approach to estimating the strengths and weaknesses of alternatives that organizations might deploy to secure their digital assets. This involves quantifying in financial terms the benefits and costs of a security measure, thereby aiding decision-makers in determining the best course of action based on economic efficiencies.

1. **Evaluating Direct and Indirect Costs**
   - **Direct Costs** include initial capital outlay for security technologies, ongoing operational expenses, and training costs for personnel.
   - **Indirect Costs** might encompass the impact on user productivity due to system downtimes, the reputational damage from potential data breaches, and the legal liabilities incurred from failing to protect sensitive data.

2. Assessing Benefits
    - **Prevention of Financial Losses:** This includes potential savings from avoiding data breaches, intellectual property theft, or any form of cyber extortion.
    - **Compliance and Reputation:** Complying with regulatory requirements can avoid hefty fines and preserve corporate reputation, thereby indirectly boosting customer trust and market value.

Implementing CBA in cybersecurity

To effectively integrate CBA into cybersecurity strategies, organizations should adopt the following methodologies:

- **Quantitative and Qualitative Measures:** While some aspects of cybersecurity benefits and costs can be quantified, such as downtime costs or compliance fines, other factors, such as brand loyalty and customer trust, require qualitative assessments. Organizations should develop a hybrid model that incorporates both to capture a comprehensive view of cybersecurity investments.
- **Risk Assessment Integration:** CBA should be aligned with thorough risk assessments to accurately identify potential vulnerabilities and the likelihood of various cybersecurity incidents. This alignment ensures that the cost-benefit calculations are grounded in realistic and relevant scenarios.
- **Dynamic Review Process:** Cybersecurity threats evolve rapidly, and so should the strategies to counteract them. Regularly updating the CBA to reflect changes in the threat landscape, technological advancements, and organizational growth is crucial to maintaining its relevance and effectiveness.

Economic theory in decision-making

Incorporating CBA into cybersecurity investment decisions provides a robust framework that supports strategic financial planning. By understanding the economic impacts of cybersecurity measures, organizations can:

- **Prioritize Investments:** Determine which cybersecurity initiatives should receive funding based on their potential to reduce risk and contribute to the organization's overall security posture.
- **Optimize Resource Allocation:** Allocate resources more efficiently by focusing on security measures that offer the highest return on investment, considering both the costs of implementation and the potential benefits in terms of risk reduction.
- **Enhance Stakeholder Communication:** Clear, quantifiable data from CBAs can help communicate the value of cybersecurity investments to stakeholders, including boards, investors, and regulators, facilitating informed decision-making and support for strategic initiatives.

Conclusion: Balancing cost and security

The integration of CBA into cybersecurity strategy enables organizations to make informed, strategic decisions regarding their investments in security measures. By carefully weighing the costs and benefits, organizations can not only ensure that their cybersecurity measures are cost-effective but also that they are sufficiently robust to protect against current and emerging threats. This economic perspective, coupled with the strategic insights provided by game theory and the behavioral insights of economics, forms a comprehensive approach to cybersecurity investment, ensuring that organizations can navigate the financial aspects of cybersecurity with confidence and precision.

SECTION 3: CORE SECURITY PRINCIPLES AND THE SIMPLIFIED CAS MODEL

Integration of core security principles

In the ever-evolving landscape of cybersecurity, the integration of core security principles within the Simplified CAS Model provides a robust framework for responding to and mitigating complex and chaotic cyber threats. This section explores how foundational security concepts – hygiene, threat detection, and mitigation – are essential building blocks within the CAS framework, facilitating dynamic and adaptable cybersecurity strategies.

Understanding core security principles

1. **Security Hygiene**
   - **Definition:** Security hygiene encompasses the basic practices and procedures designed to maintain the overall health and security of information systems. These practices are akin to routine health checks in the medical field.
   - **Implementation:** Regular updates and patch management, effective password policies, and the minimization of unnecessary services on network devices are all examples of good security hygiene. In the Simplified CAS Model, these practices form the baseline defense against potential vulnerabilities.
2. **Threat Detection**
   - **Definition:** Threat detection involves the identification and reporting of threats to prevent them from exploiting vulnerabilities in the cybersecurity infrastructure.
   - **Implementation:** Integration into the Simplified CAS Model requires sophisticated monitoring tools and anomaly detection systems that can adapt and learn from ongoing threats, thus providing real-time security insights.
3. **Mitigation**
   - **Definition:** Mitigation refers to the steps taken to reduce the severity or impact of a cyberattack once a threat is detected.
   - **Implementation:** In the context of the Simplified CAS Model, mitigation strategies involve not only immediate responsive actions but also long-term planning and adaptation strategies to prevent similar incidents in the future.

Integration within the Simplified CAS Model

The Simplified CAS Model emphasizes adaptability and responsiveness, making it particularly suitable for integrating core security principles in an environment characterized by complex and chaotic threats. Here's how these principles are woven into the fabric of the CAS framework:

- **Adaptive Security Architecture:** By embedding security hygiene, threat detection, and mitigation principles into an adaptive security architecture, organizations can ensure that their cybersecurity measures are not static. This adaptability is crucial for responding to the dynamic nature of cyber threats, where attackers continuously evolve their tactics.

- **Feedback Loops:** The Simplified CAS Model incorporates continuous feedback loops that allow systems to learn and evolve based on the latest threat intelligence and security incidents. This learning mechanism is critical for refining threat detection algorithms and enhancing mitigation techniques.
- **Proactive and Reactive Balance:** Integrating these core principles ensures a balance between proactive measures (like security hygiene) and reactive measures (like threat detection and mitigation). This balance is crucial for maintaining resilience against both anticipated and unforeseen cyber threats.

Practical application of integrated principles

The application of integrated core security principles within the Simplified CAS Model manifests in various operational scenarios:

- **Complex Threat Scenarios:** In complex environments where threats can arise unpredictably, the integration of robust threat detection systems that can adapt and respond in real time is vital. The Simplified CAS Model, supported by core security principles, enables organizations to deploy defensive mechanisms that are as dynamic as the threats they face.
- **Chaos Management:** During chaotic security incidents, such as zero-day attacks, the principles of rapid threat detection and immediate mitigation are critical. The Simplified CAS Model facilitates a swift and flexible response, minimizing potential damage and restoring system integrity.
- **Holistic Security Posture:** By aligning these core principles with the overarching goals of the Simplified CAS Model, organizations can develop a holistic security posture that not only defends against current threats but also anticipates future vulnerabilities.

Conclusion

The integration of core security principles within the Simplified CAS Model provides a strategic advantage in managing cybersecurity in a complex and unpredictable digital world. By grounding adaptive responses in foundational security practices, organizations can ensure that their cybersecurity measures are both effective and resilient, capable of evolving in step with the shifting landscape of cyber threats. This approach not only enhances immediate defensive capabilities but also supports the long-term sustainability of cybersecurity strategies.

UTILIZING THE SIMPLIFIED CAS MODEL

The Simplified CAS Model plays a pivotal role in navigating the intricate and often unpredictable realm of cybersecurity. By enabling an organization to adapt its security measures dynamically in response to emerging threats, the Simplified CAS Model is a vital tool for dealing with complex and chaotic cyber environments. This section explores the practical application of this model, emphasizing its relevance in adapting to complex and chaotic threats as categorized by the Cynefin Framework.

Adapting to complex threats

1. **Complex Threat Dynamics**
    - **Definition:** Complex threats are characterized by their unpredictability and the intricate interactions of multiple factors and agents. These threats are often difficult to foresee and require a nuanced approach to defense.

- **Application of CAS:** The Simplified CAS Model facilitates a dynamic response where security systems are not only reactive but also proactive. It incorporates a level of intelligence that allows systems to learn from each interaction and adapt their defenses accordingly.

2. **Real-Time Adaptation**
   - **Implementation:** Utilizing machine learning algorithms and artificial intelligence, the Simplified CAS Model can analyze patterns and anomalies in data to anticipate potential threats before they manifest. This proactive approach is crucial in complex threat landscapes where traditional, static security measures fall short.
   - **Example:** An AI-driven security system that dynamically adjusts firewall rules and intrusion detection systems based on the flow of network traffic and identified threat patterns.

Addressing chaotic threats

1. **Chaotic Threat Response**
   - **Definition:** Chaotic threats involve scenarios where the threat dynamics are highly unstable and unpredictable, often resulting in severe impacts.
   - **Application of CAS:** In chaotic situations, the Simplified CAS Model allows for immediate and flexible response mechanisms, which are crucial to mitigate the damage and stabilize the system.
2. **Rapid Deployment of Countermeasures**
   - **Implementation:** The model supports the deployment of automated security protocols that can instantaneously react to a detected threat, significantly reducing response times and potential damage.
   - **Example:** Deployment of emergency patches or isolation of compromised systems without human intervention during a ransomware attack detected by behavioral analytics.

Integration with the Cynefin Framework

The Simplified CAS Model and the Cynefin Framework complement each other by providing a structured approach to decision-making based on the nature of the threat environment:

- **Complex and Chaotic Threats:** Both models emphasize the need for agility and adaptability in response strategies. The Cynefin Framework helps in categorizing the threats, which in turn informs the adaptive responses facilitated by the CAS Model.
- **Aligning Responses:** By aligning the response mechanisms with the specific characteristics of each threat category defined by the Cynefin Framework, the Simplified CAS Model ensures that the security measures are both appropriate and effective.

Automating security responses

The utilization of the Simplified CAS Model extends into the realm of automated security responses, where the system not only detects threats but also responds autonomously:

- **Automated Adjustments:** Based on continuous monitoring and threat assessment, the system can autonomously adjust security controls to better protect against identified threats.

- **Feedback Loops:** Automated feedback loops integrated within the model allow for continuous improvement of the security posture based on outcomes of previous incidents and ongoing threat analysis.

Conclusion

The utilization of the Simplified CAS Model within the framework of cybersecurity provides a robust mechanism for adapting to the dynamic nature of cyber threats. By enabling real time, adaptive, and often autonomous responses to complex and chaotic threats, the model empowers organizations to maintain resilience in an ever-evolving digital threat landscape. Integrating this model with strategic frameworks like the Cynefin Framework enhances the organization's ability to categorize, understand, and react to different types of cyber threats effectively and efficiently. This approach not only mitigates immediate risks but also strengthens the overall security posture for future challenges.

SECTION 4: BENEFITS OF ADOPTING THE INTEGRATED CYBERSECURITY COMPASS (ICC)

The adoption of the ICC offers a multitude of benefits, strategically positioning organizations to navigate the tumultuous seas of the digital age with greater assurance and efficacy. This section outlines the key advantages that organizations can reap from implementing the principles and practices encapsulated in the ICC.

Enhanced decision-making

Clarity and Precision: The ICC provides a structured approach to decision-making, leveraging models like the Cynefin Framework and the Simplified CAS Model. These tools help classify threats and determine the most effective response strategies based on the nature of the threat. This structured approach reduces ambiguity and enhances the precision of security measures.

Example: Using the Cynefin Framework, a company can quickly categorize an emerging threat as "complex" and deploy AI-driven adaptive security measures that are best suited for such unpredictable scenarios.

Proactive security posture

Anticipation of Threats: By integrating advanced predictive models and continuous monitoring, the ICC enables organizations to anticipate and prepare for potential threats before they manifest. This shift from a reactive to a proactive security stance significantly reduces potential damages and enhances the organization's resilience.

Example: Real-time data analytics can predict potential security breaches by identifying patterns that precede an attack, allowing preemptive action to be taken.

Dynamic adaptability

Agility in Response: The ICC promotes an adaptive security environment where systems and policies are continuously refined based on the latest data and emerging threats. This adaptability is crucial in a landscape where threat actors constantly evolve their tactics.

Example: An organization might use the Simplified CAS Model to automatically adjust its encryption levels in response to detected increases in network sniffing activities.

Comprehensive risk management

Holistic Approach: The ICC encourages a holistic view of cybersecurity, considering not only technical factors but also economic, behavioral, and strategic aspects. This comprehensive approach ensures that all potential impacts of cybersecurity decisions are considered, leading to more robust risk management.

Example: Incorporating economic theory into cybersecurity strategy allows an organization to allocate its resources more efficiently, focusing on areas with the highest potential return on security investment.

Improved compliance and alignment with regulations

Regulatory Adherence: With its emphasis on evidence-based practices and structured frameworks, the ICC helps organizations meet compliance requirements more effectively. The clear and documented process and decision trails facilitated by the ICC also simplify the auditing process.

Example: The detailed logging of decision-making processes and the data supporting those decisions can provide auditors with the necessary information to verify compliance with General Data Protection Regulation (GDPR) or other regulatory frameworks.

Cultural shift toward security

Security as a Shared Responsibility: By fostering an environment where security is seen as a collective responsibility, the ICC helps to cultivate a culture of security within the organization. This cultural shift is supported by the continuous education and involvement of all stakeholders in the security processes.

Example: Regular training sessions and simulations can help employees understand their role in the organization's cybersecurity efforts, making them more likely to adhere to security policies and procedures.

Cost efficiency

Optimized Resource Utilization: The strategic application of resources guided by the ICC ensures that investments in cybersecurity are both effective and efficient. The use of cost-benefit analyses and game theory helps in prioritizing investments that yield the highest protective returns.

Example: By analyzing the cost of potential data breaches against the cost of implementing advanced security measures, organizations can better manage their budgets, focusing spending on high-impact solutions.

Conclusion

The adoption of the ICC equips organizations with the tools, knowledge, and strategies to confidently face the challenges of the digital era. The benefits extend beyond enhanced security measures, fostering a culture of resilience, proactive risk management, and regulatory compliance. As organizations continue to grapple with the complexities of cybersecurity, the ICC offers a navigational chart that not only addresses current security needs but also prepares them for future challenges.

CONCLUSION

As we conclude our exploration of the ICC, it is evident that this comprehensive framework serves as a robust navigational tool, expertly designed to guide organizations through the intricate and often perilous cybersecurity landscape. The ICC's multifaceted approach, which incorporates the Cynefin Framework, Resilience Engineering, economic theories, and the Simplified CAS Model, provides a strategic roadmap that ensures a holistic and adaptable security posture.

Recap of the ICC's role

The ICC empowers organizations to effectively categorize and respond to a wide range of cyber threats with agility and informed precision. By embedding core security principles into a dynamic, principle-based structure, the ICC enables a transition from reactive security measures to a proactive, anticipatory stance. This shift is crucial in a world where cyber threats are not only growing in number but also in sophistication.

Importance of a principle-based approach

The principle-based approach championed by the ICC underscores the necessity of grounding cybersecurity strategies in solid, time-tested foundations that are adaptable to the changing dynamics of threat landscapes. This approach ensures that cybersecurity measures are not merely reactive or superficial but are deeply rooted in empirical evidence, strategic analysis, and continuous improvement. Such a foundation is indispensable for enduring the challenges that arise in an environment marked by rapid technological evolution and increasingly sophisticated cyber threats.

Encouragement for future adoption

For organizations looking toward the future, the adoption of the ICC offers more than just security; it provides a strategic advantage. By embracing the ICC, organizations can ensure that their cybersecurity measures are as dynamic and sophisticated as the threats they face. This proactive and informed approach is critical for navigating future uncertainties with confidence.

As the digital frontier continues to expand and evolve, the principles embedded in the ICC will serve as a compass, guiding organizations through uncharted waters. The journey of cybersecurity is ongoing, and while the threats may evolve, the foundational strategies provided by the ICC will continue to empower organizations to protect their digital landscapes effectively.

In conclusion, the adoption of the ICC is not just a strategic decision for enhanced cybersecurity but a commitment to a resilient, informed, and proactive approach to navigating the complexities of the digital age. Organizations that embrace this comprehensive and principle-based framework will be well equipped to face whatever challenges the future may hold, ensuring their voyage through the cybersecurity odyssey is both secure and successful.

CYBER CHUCKLES: NAVIGATING THE ICC WITH A GRIN

Welcome aboard the S.S. Cyber Secure, where navigating the treacherous waters of cybersecurity isn't just about staying afloat – it's about enjoying the voyage! Let's crack open the treasure chest of humor with a light-hearted recap of our journey through the ICC. Grab your lifejackets, and let's dive in!

1. **Weathering the Storm with the Cynefin Framework**
   - **Joke:** Why did the cybersecurity team consult their Cynefin Framework during a hurricane? Because when things get chaotic, it's the only thing that doesn't lose its bearings!
   - **Motto:** "When in doubt, chart it out!"
2. **Resilience Engineering: More Than Just Duct Tape**
   - **Joke:** Why do cybersecurity engineers love caffeine? Because it helps them "stay awake" to patch system vulnerabilities before they crash the network party!
   - **Motto:** "Keep calm and resilience on!"
3. **Economic Theories in Cybersecurity – Because Money Matters**
   - **Joke:** Why did the cybersecurity expert study economics? To learn why spending a penny on prevention is better than a pound on a cure!
   - **Motto:** "Save a dime, secure in time!"
4. **Game Theory: Predicting the Hacker's Next Move**
   - **Joke:** Why do cybersecurity strategists play chess? To checkmate hackers in their own game – without losing their kings in the process!
   - **Motto:** "Think like a hacker, act like a keeper!"
5. **Cost-Benefit Analysis: Not Just for Accountants**
   - **Joke:** How many cybersecurity experts does it take to perform a CBA? One to calculate the risks, and another to explain why it's still cheaper than a data breach!
   - **Motto:** "Measure twice, cut cyber threats once!"
6. **Simplified CAS Model: Keeping It Simple, Cyber Sailor**
   - **Joke:** Why do cybersecurity teams use the Simplified CAS Model? Because even in cyber seas, simplicity sails smoother!
   - **Motto:** "Complexity out, clarity in!"
7. **Benefits of the ICC: Smooth Sailing Ahead:**
   - **Joke:** What's a hacker's least favorite thing about the ICC? It makes cybersecurity less about guessing and more about outclassing!
   - **Motto:** "With the ICC, we see clearer seas!"

As we dock at the end of this chapter, remember that while the waves of cybersecurity challenges may be relentless, armed with the ICC and a sense of humor, you're well equipped to navigate through. And if all else fails, just remember: the best cybersecurity strategy starts with a good laugh and a better backup!

Cyber Cheers to Your Adventures on the Digital High Seas!

Chapter 24

The last chapter

Embracing a principle-based philosophy in the cybersecurity odyssey

Spoiler Alert!

- Ahoy, mateys! This final chapter marks the end of your cybersecurity journey.
- Set sail on a sea of philosophical principles to guide you through the ever-changing waters of cybersecurity.
- This chapter emphasizes the importance of basing your security decisions on sound evidence and logic (think using a compass and sextant instead of just following rumors of buried treasure).
- Learn the power of "Occam's Razor" – the principle that the simplest explanation is usually the right one (because sometimes fancy solutions can make things worse).
- This chapter doesn't offer a happily-ever-after ending, because cybersecurity is an ongoing battle.
- Instead, it equips you with the tools and philosophy to face future threats with confidence (because a true cybersecurity warrior is always prepared).

NAVIGATING THE CYBERSECURITY ODYSSEY WITH PRINCIPLE-BASED PHILOSOPHIES

The digital era unfolds as a vast, ever-changing landscape. Here, cyber threats proliferate with each technological advancement, casting long shadows over the domain of bytes and data. In this perpetual struggle between security measures and emerging threats, the cybersecurity community stands vigilant, adapting and responding with unwavering dedication. Within this dynamic context, the necessity for a principle-based philosophy in cybersecurity becomes strikingly clear – a foundational ethos built not just on the latest technologies but on enduring principles that guide actions and strategies.

The concept of adopting a principle-based philosophy for navigating the complexities of cybersecurity highlights the indispensable role of core tenets: empirical evidence, progress-oriented discussions, and the guiding principle of simplicity. These pillars serve not only as navigational beacons but also as the very foundation upon which cybersecurity defenses are constructed. This chapter unfolds the argument for a principled approach to cybersecurity, illuminating how such a philosophy can ground actions in wisdom and strategies in evidence, offering a resilient framework in the face of digital adversities.

DOI: 10.1201/9781003562184-31

EMPIRICAL FOUNDATION: THE CORNERSTONE OF EFFECTIVE CYBERSECURITY

The uncertain realm of cybersecurity, teeming with conjectures and evolving threats, demands a steadfast reliance on empirical evidence. This commitment to evidence-based practices ensures that strategies and protective measures are not constructed on mere speculation but are informed by real-world data and experiences. Empirical evidence, drawn from case studies, incident reports, and rigorous data analysis, offers a beacon in the fog, shedding light on vulnerabilities, adversarial tactics, and the efficacy of various security measures.

Grounding cybersecurity practices in empirical evidence does more than just illuminate the path forward; it fortifies defenses with the weight of proven insights, ensuring approaches to safeguarding digital assets are both informed and resilient. The cornerstone of effective cybersecurity lies not in the algorithms and firewalls alone but in the empirical data that shapes and refines these tools. In a realm rife with uncertainties, empirical evidence provides a measure of certainty, a solid foundation upon which to base decisions and strategies in the ever-evolving landscape of digital threats.

PROGRESS-ORIENTED DISCUSSIONS: PAVING THE PATH TOWARD RESILIENCE

The journey toward cybersecurity resilience is a collaborative endeavor, marked by continuous learning, adaptation, and collective effort. It is through progress-oriented discussions that the cybersecurity community navigates the challenges inherent in protecting the digital domain. This chapter champions the pursuit of practical advancements through dialogue, stressing the importance of constructive conversations in overcoming obstacles and expanding the horizons of collective knowledge.

Collaborative efforts and the sharing of wisdom form the lifeblood of progress in the field of cybersecurity. By fostering an environment where ideas are exchanged freely and diverse perspectives are valued, a dynamic ecosystem emerges, better equipped to respond to and anticipate challenges. Such progress-oriented discussions not only advance collective knowledge but also strengthen the fabric of digital defenses, uniting the community in a shared mission to push forward into uncharted territories, fortified by a resolve to achieve tangible progress.

SIMPLICITY AS A GUIDING PRINCIPLE: OCCAM'S RAZOR AND BAYESIAN THINKING

Amidst the intricate tapestry of cybersecurity decision-making, simplicity emerges as a guiding principle, cutting through complexity with clarity and precision. Drawing on the philosophical tenets of Occam's Razor, complemented by Bayesian thinking, this principle advocates for favoring straightforward solutions amid a plethora of potential paths. Embracing simplicity in the design of security systems and policies enhances their clarity, manageability, and overall efficacy.

Simplicity, in the context of cybersecurity, does not denote a lack of sophistication but rather an elegance in distilling complex problems to their essence, allowing for solutions that are effective and adaptable. This adherence to simplicity ensures that cybersecurity defenses are not only robust but also agile, capable of evolving in response to the shifting landscape of digital threats.

THE FINISHING LINE

As this exploration of cybersecurity's principle-based philosophy concludes, it's essential to reflect on the journey that has unfolded across these pages. From the foundational reliance on empirical evidence to the collaborative spirit of progress-oriented discussions, and the clarity brought forth by simplicity, each chapter has woven together a tapestry of principles that, collectively, fortify our digital defenses in an age of relentless cyber threats.

This book has endeavored to guide readers through the nuanced landscape of cybersecurity, emphasizing that the strength of our defenses lies not solely in the sophistication of our technologies but in the depth of our understanding and the steadfastness of our principles. As we navigate the complexities of protecting digital assets, it is these principles that will light our way, providing a beacon amidst the ever-evolving challenges of the digital frontier.

In closing, let this work serve as both a compass and a call to arms for cybersecurity professionals, policymakers, and enthusiasts alike. The path forward is one of continuous learning, adaptation, and principled action. By embracing the philosophical underpinnings discussed herein, we can forge a future where cybersecurity is not just a reactive measure but a foundational aspect of our digital society, imbued with the wisdom and insight necessary to anticipate and neutralize threats before they arise.

May the principles explored in this book inspire you to contribute to the collective resilience and security of our digital world. Together, armed with knowledge, wisdom, and an unwavering commitment to our core tenets, we stand ready to face the cybersecurity challenges of today and tomorrow.

With that, the journey does not end but rather evolves into the next chapter of our continuous quest for security in the digital age. Thank you for embarking on this journey through the principles of cybersecurity, and may your path be enlightened by the insights and philosophies shared within these pages.

CYBER CHUCKLES: EMBRACING A PRINCIPLE-BASED PHILOSOPHY IN THE CYBERSECURITY ODYSSEY – A SWASHBUCKLING GUIDE

Ahoy, digital sea-farers and cyber-scurvy dogs! As we draw the curtains on this grand cybersecurity odyssey, we hoist the sails for one final voyage – a journey not of land or sea, but through the tempestuous waves of principles and philosophies. Here, in the treacherous waters of the cyber realm, our compass and sextant are not tools of navigation but beacons of wisdom, guiding us through storms and squalls with sound evidence and steadfast logic.

Charting the course with empirical stars

In the starlit sky of cybersecurity, empirical evidence shines brightest, guiding our ship away from the rocky shores of speculation. Like seasoned navigators interpreting celestial patterns, we align our security measures with the constellations of data, steering clear of mythical sea monsters whispered in the tales of less empirical sailors. Our decisions are anchored in the solid ground of case studies and data analysis, ensuring our voyage is informed by the gleaming stars of proven insights.

Dialogue: The wind in our sails

Our journey is propelled forward by the winds of progress-oriented discussions, where every voice at the helm is heard, from the seasoned captain to the greenest deckhand. In the great hall of our ship, tales of cyber battles won and lost are shared, not for glory, but for learning. It's here, amidst the clinking of mugs and the roar of tales, that our collective wisdom grows, charting a course toward resilience and unity against the digital tempests ahead.

Simplicity: Our trusted cutlass

In the arsenal of cybersecurity, simplicity is our cutlass – sharp, reliable, and swift in the hands of a skilled navigator. Guided by the wisdom of Occam's Razor, we discover that the simplest strategies often steer us through the most turbulent cyber maelstroms. Our policies and defenses, shaped by the principle of simplicity, are as sleek and agile as our fastest ship, prepared to adjust to the shifting winds of cyber threats with precision and grace.

The voyage ever continues

As we stand at the helm, gazing out at the horizon where the digital sky meets the sea of information, we understand that our journey never truly ends. The cybersecurity odyssey is one of eternal vigilance, a perpetual battle against the sirens and leviathans that lurk in the depths of the cyber ocean. But fear not, for we are armed with the mightiest of weapons: a principle-based philosophy that guides our every action, decision, and strategy.

A toast to the future

So, here's to you, brave voyagers of the cyber seas. May your compass always be true, your cutlass ever sharp, and your sails filled with the winds of progress and dialogue. Though the seas may be uncertain and the journey long, remember – the true treasure lies not in gold or jewels but in the wisdom gained and the battles bravely fought.

Raise your tankards high, for though this chapter may close, the odyssey continues. Onward, to uncharted waters, guided by the stars of principles and the steadfast heart of a true cybersecurity warrior. Here's to the future, to resilience, and to the endless pursuit of securing our digital world.

AND WITH THAT, WE SET OUR SIGHTS ON NEW ADVENTURES, EVER READY TO FACE THE CHALLENGES THAT LIE AHEAD, WITH A PRINCIPLE-BASED PHILOSOPHY AS OUR GUIDING STAR.

Index

www.ingramcontent.com/pod-product-compliance
Ingram Content Group UK Ltd.
Pitfield, Milton Keynes, MK11 3LW, UK
UKHW052051230926
475702UK00025B/1113
* 9 7 8 1 0 3 2 9 1 0 3 7 6 *